National Academy Press

The National Academy Press was created by the National Academy of
Sciences to publish the reports issued by the Academy and by the
National Academy of Engineering, the Institute of Medicine, and the
National Research Council, all operating under the charter granted to
the National Academy of Sciences by the Congress of the United States.

Prudent Practices for Disposal of Chemicals from Laboratories

Committee on Hazardous Substances
in the Laboratory

Commission on Physical Sciences, Mathematics,
and Resources

National Research Council

NATIONAL ACADEMY PRESS
Washington, D.C.   1983

National Academy Press, 2101 Constitution Avenue, NW, Washington, DC 20418

NOTICE: The project that is the subject of this report was approved by the Governing Board of the National Research Council, whose members are drawn from the councils of the National Academy of Sciences, the National Academy of Engineering, and the Institute of Medicine. The members of the committee responsible for this report were chosen for their special competences and with regard for appropriate balance.

This report has been reviewed by a group other than the authors according to procedures approved by a Report Review Committee consisting of members of the National Academy of Sciences, the National Academy of Engineering, and the Institute of Medicine.

The National Research Council was established by the National Academy of Sciences in 1916 to associate the broad community of science and technology with the Academy's purposes of furthering knowledge and of advising the federal government. The Council operates in accordance with general policies determined by the Academy under the authority of its congressional charter of 1863, which establishes the Academy as a private, nonprofit, self-governing membership corporation. The Council has become the principal operating agency of both the National Academy of Sciences and the National Academy of Engineering in the conduct of their services to the government, the public, and the scientific and engineering communities. It is administered jointly by both Academies and the Institute of Medicine. The National Academy of Engineering and the Institute of Medicine were established in 1964 and 1970, respectively, under the charter of the National Academy of Sciences.

This material is based on work supported by the National Institutes of Health, the National Science Foundation, and the Environmental Protection Agency under Grant No. PRM-8120234, Amendment No. 1 and Amendment No. 2; the American Chemical Society; and the Chemical Manufacturers Association.

Library of Congress Cataloging in Publication Data

Main entry under title:

Prudent practices for disposal of chemicals from
  laboratories.

  Bibliography: p.
  Includes index.
  1. Chemical laboratories—Waste disposal.
I. National Research Council (U.S.). Committee on
Hazardous Substances in the Laboratory.
QD51.P77 1983   542'.028'9   83-13211
ISBN 0-309-03390-X

Printed in the United States of America

Subcommittee on Wastes from Biomedical Laboratories

BLAINE C. McKUSICK, Wilmington, Delaware
EDWIN D. BECKER, National Institutes of Health
IRVING H. GOLDBERG, Harvard Medical School

Subcommittee on Disposal Procedures for Specific Laboratory Chemicals

BLAINE C. McKUSICK, Wilmington, Delaware
IRVIN L. KLUNDT, Aldrich Chemical Company
KENNETH L. WILLIAMSON, Mount Holyoke College

Subcommittee on Activities by Other Organizations (including Europe)

P. CHRISTIAN VOGEL, BASF Wyandotte Corporation
ALAIN DeCLEVE, Stanford University

Commission on Physical Sciences, Mathematics, and Resources

GERHART FRIEDLANDER, Brookhaven National Laboratory

Commission on Life Sciences

RONALD W. ESTABROOK, University of Texas Health Science Center

NRC Staff

WILLIAM SPINDEL, *Study Director*
ROBERT M. JOYCE, *Consultant*
PEGGY J. POSEY, *Staff Associate*
WENDY L. BAKER, *Administrative Secretary*
NANCY H. DOYLE, *Secretary*

v

vii

Preface

Prudent Practices for Handling Hazardous Chemicals in Laboratories,[*]
published under the aegis of the National Academy of Sciences, includes
brief sections on disposal of chemicals and wastes from laboratories
(Chapter II, Sections E, F, and G). During the latter stage of the preparation of that report, the initial regulations on management of hazardous waste, prepared by the U.S. Environmental Protection Agency (EPA)
under the authority of the Resource Conservation and Recovery Act
(RCRA), were made effective (19 November 1980). In June 1981 an
ad hoc planning group organized by the Assembly of Mathematical and
Physical Sciences of the National Research Council (NRC) concluded
that many laboratories—academic, government, and industrial—were
having substantial problems in understanding and complying with these
complex regulations. This group recommended that the NRC undertake
a study of the disposal of chemicals from laboratories, to deal with the
subject in greater detail, by assessing the impact of the EPA RCRA
regulations on laboratories, by developing guidelines for laboratories in
the disposal of chemicals, and by recommending changes in laboratory
procedures and in regulations that would simplify the disposal of chemicals from laboratories in ways that would be safe and environmentally
acceptable.

The procedures and recommendations were to be framed for laboratory operations, where many types of chemicals are used in relatively

small quantities, and were not to be generally applicable to operations such as manufacturing, where chemicals are used on a large scale.

Funding for the study was obtained from private and government sources, and a parent committee and eight subcommittees broadly representative of academic, government, and industrial laboratories, both large and small, were constituted. The experience of the individuals on these committees is reflected in the guidelines and recommendations in this report. Moreover, to get a broader range of opinion from laboratory workers and managers, for whom the report is written, draft copies were distributed to a number of organizations and individuals concerned with disposal of chemicals and waste from laboratories. Their comments and suggestions have been considered in preparing the final report.

The committee is pleased to acknowledge assistance from many people who made important contributions to the study, but the committee is solely responsible for any errors in the report. The study was initiated while Theodore L. Cairns was Chairman of the NRC Office of Chemistry and Chemical Technology, and he provided broad guidance. Special note should be made of the contributions of Robert M. Joyce, technical consultant, who, in addition to his technical input, organized the material from the various subcommittees and put the report in its final form. The committee is also indebted to William C. Drinkard, who wrote the sections on disposal of inorganic and organo-inorganic chemicals in Chapter 6. William Spindel, Executive Secretary of the Board on Chemical Sciences and Technology, made many contributions to the development and execution of the study. We would also like to acknowledge the contributions made by Peggy J. Posey, Wendy L. Baker, and the typist, Nancy H. Doyle.

> ROBERT A. ALBERTY, *Chairman*
> Committee on Hazardous Substances
> in the Laboratory

*NRC Committee on Hazardous Substances in the Laboratory, *Prudent Practices for Handling Hazardous Chemicals in Laboratories*, National Academy Press, Washington, D.C., 1981.

Contents

xi

Prudent Practices for Disposal of Chemicals from Laboratories

Overview and Recommendations

INTRODUCTION

All laboratory work with chemicals eventually produces chemical waste, and those who generate such waste have moral and legal obligations to see that the waste is handled and disposed of in ways that pose minimum potential harm, both short term and long term, to health and the environment. The legal obligations began on 19 November 1980 when the U.S. Environmental Protection Agency (EPA) put into effect federal regulations on a Hazardous Waste Management System,[1] under the authority of the Resource Conservation and Recovery Act (RCRA) of 1976, as amended. These regulations are designed to establish a "cradle-to-grave" system for the management of hazardous wastes from all sources.

The objectives of this report are to present guidelines for laboratories in establishing a waste management system, to give specific recommendations to laboratory managers for the disposal of chemicals, and to make recommendations for constructive changes in regulations for disposal of chemicals from laboratories.

For the purposes of this report, a laboratory is defined as a building or area of a building used by scientists or engineers, or by students or technicians under their supervision, for the following purposes: investigation of physical, chemical, or biological properties of substances;

1

development of new or improved chemical processes, products, or applications; analysis, testing, or quality control; or instruction and practice in a natural science or in engineering. These operations are characterized by the use of a relatively large and variable number of chemicals on a scale in which the containers used for reactions, transfers, and other handling of chemicals are normally small enough to be easily and safely manipulated by one person.

Broadly, a hazardous chemical is a chemical that poses a danger to human health or the environment if improperly handled. The EPA RCRA regulations define hazardous waste, for regulatory purposes, in terms of specific hazard characteristics and by listing specific chemicals and residues from chemical operations. Even though many of the chemicals synthesized or used in laboratories do not meet any of the EPA regulatory criteria, they must be considered hazardous because of unknown toxicity. This report is concerned with all laboratory chemicals that must be handled and disposed of as hazardous regardless of whether they are covered by government regulations. Biological waste generated by life-science laboratories is covered only to the extent that the biological waste is contaminated with chemicals, for example, waste generated in animal or microbiological tests of toxicity or mode of action of chemicals. This report does not cover waste that is regulated by other agencies under other laws, for example, radioactive waste by the Nuclear Regulatory Commission.

The principal direction of the EPA RCRA regulations is toward industry, which generates the preponderance of hazardous waste in the United States. However, the regulations apply equally to laboratories, which are estimated by EPA to generate only 0.1–1% of the total U.S. hazardous waste. Although the hazard inherent in a small quantity of a chemical from a laboratory is the same as the hazard inherent in a much larger quantity of the same chemical from another source, the overall potential for harm to health or the environment is less from the former because of the smaller quantity. A large fraction of laboratory waste comprises small amounts of many kinds of chemicals; such waste is generally sent to a secure landfill, sometimes hundreds of miles from the laboratory, because many operators of commercial incinerators do not accept these low-volume, chemically diverse wastes. Because of these constraints on disposal, and the associated extensive recordkeeping required, many of the nation's laboratories are having problems in setting up effective systems for handling their hazardous waste and, in some cases, in finding legal and safe ways for disposing of it.

Many of the recommendations made in this report are based on the experience of laboratory managers and personnel who have implemented effective chemical waste management programs. The recommendations are based on successful application of scientific concepts that reduce or eliminate negative environmental impacts of laboratory waste disposal. Many of these concepts have been known and utilized for years prior to the hazardous-waste regulations. However, some of them may not be strictly in accord with current regulations, and it is important that everyone comply with all regulations in force at any given time. These recommendations are presented with the objective of conveying concepts and methodologies that work.

OVERVIEW AND SUMMARY

The primary purpose of this report is to give guidance to the laboratory manager, scientist, engineer, student, and technician on prudent practices for disposal of chemicals.

Chapter 1 outlines the characteristics and requirements of a waste management system. It emphasizes the need for executive commitment to the program, for written procedures, and for assigned responsibilities.

Figure 1 presents a guide to the various methods of chemical disposal discussed in this report. The advent of the EPA RCRA regulations has stimulated research and development on methods of chemical disposal other than incineration or landfilling. These newer processes, such as chemical destruction in supercritical water or molten salt baths, are not discussed in this report because they have not yet been developed to the point of offering practical options to laboratories for disposal of laboratory chemicals.

A laboratory worker faced with unneeded material must provide information on the properties of the material to guide waste management personnel in selecting the method of disposal. The first option that should be considered is whether the material can be reused, recycled, or recovered for reuse; these options are discussed in Chapter 4.

If it is decided that the material is a waste, it must be determined whether it is a regulated hazardous waste, using the criteria outlined in Chapter 2. Even if it is not a regulated hazardous waste, a decision must be made as to whether to handle it as hazardous, and its route of disposal should be governed by its characteristics according to the five classes in Figure 1—combustible, noncombustible, biological, explosive, or radioactive. (This report does not deal with radioactive waste; see Chapter 11 for pertinent references.) The hazardous waste must then be properly labeled and segregated. These topics are also dealt with in Chapter 2.

Chapter 3 covers the storage of laboratory waste. The physical problems of storage—protection against the elements, fire, and leakage—as

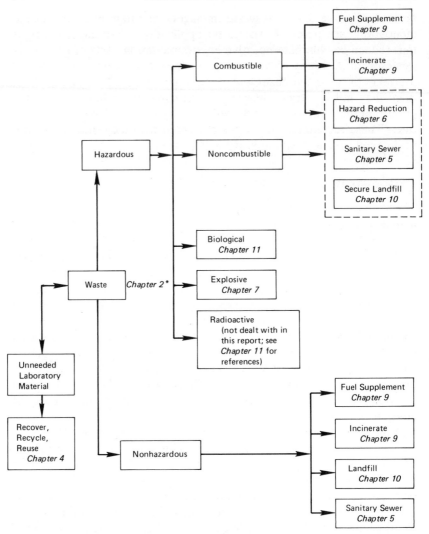

FIGURE 1 Decision tree for disposal of chemicals from laboratories. *See Chapter 2 to determine whether a waste is classified as hazardous.

well as the management aspects—inspection, recordkeeping, and contingency planning—are discussed.

After waste has been properly classified and segregated, it must eventually be disposed of in some ecologically prudent manner. If the waste is nonhazardous it can be incinerated (Chapter 9); sent to a municipal landfill (Chapter 10); or, under certain circumstances, put in the sanitary sewer (Chapter 5). Modest quantities of many common chemicals can be safely and acceptably disposed of down the drain if the precautions given in Chapter 5 are followed. The specialized problems involving the disposal of chemically contaminated waste from life-science laboratories are discussed in Chapter 11. The handling and disposal of explosives are discussed in Chapter 7.

Most hazardous waste will fall into the classes defined as combustible or noncombustible. Within these categories waste material can range from acutely hazardous to mildly hazardous. Naturally much more care and money are needed to store, transport, and dispose of acutely hazardous material. A trained professional can greatly reduce or completely destroy the hazard characteristic of many hazardous chemical wastes by chemical reaction in the laboratory. The longest chapter in this book (Chapter 6) gives detailed procedures for the destruction of laboratory quantities of common classes of hazardous organic and inorganic chemicals, particularly those that cannot be put in a landfill because of the characteristic of reactivity.

If the waste is not destroyed in the laboratory or should not be disposed of in the sewer, it must be either incinerated or buried. Since most laboratories are not equipped to carry out these operations, the waste must usually be transported to an appropriate incinerator or landfill. Chapter 8 outlines the packaging and loading of waste and the many EPA and U.S. Department of Transportation (DOT) regulations governing the transport of waste.

Although incineration destroys most hazardous waste, many commercial incinerator operators will not accept laboratory waste because of its small quantity and chemical diversity. Commercial incineration is more costly than landfill disposal. The costs of equipment, preliminary testing required by regulations, and operation preclude most laboratories from installing their own incineration equipment. Chapter 9 discusses the use and operation of on- and off-site incinerators as well as the possibility of burning combustible waste in a power or steam-generating plant.

At present, a significant fraction of laboratory waste goes to landfills. Although some laboratories make their own arrangements for transportation and off-site disposal of chemicals, others find it more economical to employ commercial firms that will pack laboratory chemicals and arrange for their transportation and disposal. Chapters 9 and 10 provide guidance on the selection of a contract waste-disposal firm. Chapter 10 lists the types of chemicals that can go to a sanitary landfill, regulations regarding secure landfills, and the preparation of waste destined for landfills.

The appendixes to this report provide detailed information to guide the disposal of laboratory chemicals. The regulations that govern transportation and disposal of waste are voluminous and complex. Appendix A contains summaries of, and copies of parts of, EPA regulations that are particularly pertinent to generators of laboratory waste. These regulations are those that were effective 1 April 1983; future changes in them will be found in subsequent issues of 40 CFR Subchapter I. Sources that can be used to keep abreast of proposed and actual changes in federal regulations are given in Appendix M. Appendix B gives the telephone numbers of the state hazardous-waste regulatory agencies as well as a tabulation of how the various states stand (1982) with regard to exemption of small quantity waste generators from EPA regulations. Appendix C covers EPA regulations on the storage of laboratory waste, and Appendix D covers DOT regulations on the transportation of hazardous chemicals.

In storing, packing, and shipping waste, one is faced with the problem of deciding which materials can be grouped together. Appendixes E–I give examples of chemically incompatible, potentially explosive, water-reactive, pyrophoric, and peroxide-forming chemicals, respectively.

Many water-soluble, inorganic cations can be precipitated as hydroxides to give low-volume solids for burial, and the pH ranges for this process are given in Appendix J. Appendix K gives guidelines for disposal of chemicals in the sanitary sewer system. Appendix L contains descriptions of incineration equipment. Finally, a glossary that includes abbreviations used in the text is provided.

RECOMMENDATIONS FOR LABORATORY MANAGERS AND PERSONNEL

The general principles that should underlie a system and procedures for managing unneeded laboratory chemicals are identical to those of an effective laboratory safety program.[2] Hazardous chemicals can be handled safely and disposed of by environmentally acceptable methods. As this committee said in an earlier report, "Chemicals must be disposed of in such a way that people, other living organisms, and the environment generally are subjected to minimal harm by the substances used or produced in the laboratory. Both the laboratory workers and the supporting personnel should know and use acceptable disposal methods for various chemicals."[2] All laboratory management should implement policies and practices to achieve these goals. The following recommendations are directed to actions that can be taken by laboratories to avoid or minimize problems associated with unneeded chemicals.

WASTE MANAGEMENT SYSTEM

The officer or manager who is responsible for a laboratory should establish and support, on a continuing basis, a laboratory waste management policy. A laboratory site that has other activities usually has an overall waste management policy, but it also needs a subsidiary policy specific to laboratory operations. The laboratory policy should be implemented through written procedures and assigned responsibilities. It is essential that laboratory management at all levels, or faculty in an academic institution, be openly and actively committed to support of sound waste management policies and practices. It is equally important that laboratory personnel who are responsible for handling and disposing of chemicals be thoroughly familiar with and comply with all applicable regulations—federal, state, and local (see Chapter 1).

REDUCTION IN VOLUME OF WASTE

Planning Experiments

The plan for every experiment should include consideration of the disposal of leftover starting materials and the products and by-products that will be generated (see Chapter 1, Section V.A). Such planning can reduce the volume of waste generated in the experiment and may also avoid difficult disposal problems.

Reduce the Scale of Experiments

One route to the reduction of the volume of hazardous waste from laboratories engaged in teaching, research, or testing is the use of modern technology to reduce the scale on which experiments are carried out (see Chapter 1, Section V.B).

Exchange or Recovery of Chemicals

Laboratories should consider possibilities for exchange of unneeded chemicals and for recovery of used solvents (see Chapter 4). These procedures can reduce the costs of purchasing new chemicals and of disposal.

PREVENTION OF SPECIAL DISPOSAL PROBLEMS

Indefinite and uncontrolled accumulation of excess reagents can create safety hazards and storage problems. These problems can be alleviated, and purchase costs saved, by instituting an excess-chemicals store to which laboratory workers can go for chemicals instead of ordering new material (see Chapter 4, Section IV). Chemicals that are prone to deteriorate with time can pose difficult disposal problems if allowed to accumulate. The waste management system (see Chapter 1) should provide a mechanism for periodic checking for such reagents.

Reactive Chemicals

Reagents that react readily with oxygen or water are prone to deteriorate when stored for long times after the original container has been opened. Disposal problems can be created by old samples of water-reactive chemicals (see Appendix G), pyrophoric chemicals (see Appendix H), and peroxide-forming chemicals (see Appendix I). Procedures for laboratory destruction of all three of these classes are given in Chapter 6.

The handling of explosive chemicals is discussed in Chapter 7.

Reagents with Deteriorated Labels

Deterioration of labels is a common occurrence on old reagent containers (see Chapter 1, Section V.D). If the reagent or container has not deteriorated, the container should be relabeled if its identity is certain. If the label has disappeared or become illegible, and the nature of the chemical is unknown, the chemical should be discarded. The method for proper disposal can usually be determined by simple laboratory tests (see Chapter 6, Section V).

Reaction Mixtures

Laboratory glassware containing reaction mixtures of unknown nature, and sometimes of unknown origin, can pose difficult disposal problems. Such materials can be a frequent occurrence in research laboratories, particularly in those that have a high rate of personnel turnover. The disposal of an unlabeled orphan reaction mixture can sometimes be guided by knowledge of the chemistry that was being done by a departed laboratory worker. Simple laboratory tests (see Chapter 6, Section V) will often provide enough information to guide safe disposal of a material. The waste management system should provide a procedure designed to prevent the occurrence of such orphan wastes (see Chapter 1, Section V.E).

EDUCATIONAL ASPECTS OF WASTE MANAGEMENT

Many of the scientists being educated in our colleges and universities today will move into professions from which society expects responsible, informed behavior in the disposal of hazardous chemical substances. The time to learn this behavior is in the high school and college or university laboratory. Educators, many of whom were trained in a more casual era, must now be responsible, by precept and example, for teaching proper methods for the disposal of hazardous laboratory waste. Teachers should demand that a formal waste-disposal program be initiated if it has not been and should see that the program is implemented in their laboratories. Undergraduate laboratories should be equipped with the necessary disposal equipment, and the proper disposal of wastes should be a part of every laboratory procedure. The authors of laboratory texts should deal with the problems of disposal explicitly and should include as part of each experiment procedures that the student can carry out to minimize disposal problems.

Some academic laboratories have formal safety programs, with assigned responsibilities to safety personnel or faculty to instruct students in safe laboratory practices and to monitor safety in the laboratory. Such programs can logically be extended to cover the equally important area of handling and disposing of waste from laboratory operations.

In some academic institutions the disposal of hazardous laboratory chemicals can be incorporated into the curriculum. Professionals trained in this area will be in demand, and therefore programs on chemical disposal from laboratories can be training programs. Chemistry students can learn about detoxifying or destroying chemicals, microbiology students can come to grips with aerobic and anaerobic destruction of toxic materials, and engineering students can study design of incinerators and other waste-handling equipment.

At least one academic institution has funded two graduate research positions through its hazardous-waste program. This step not only provides chemical support for the waste management program but also increases the interest of chemistry department faculty in hazardous-waste management.

The literature contains few good, detailed procedures for destroying pyrophoric, water-reactive, explosive, or highly toxic chemicals. Chemists could make a useful contribution to their profession by including such procedures in their publications or by submitting them for publication in journals. It may even be useful to have a publication similar to *Organic Syntheses*, where checked procedures of this kind could be published.

FUTURE DIRECTIONS

It is essential that everyone comply with the federal, state, and local regulations that are in force at any given time, and it should therefore be emphasized that the following recommendations are not in accord with current regulations. This committee believes that the handling and disposal of laboratory chemicals can be simplified without posing a hazard to health or a threat to the environment. The basis for possible simplification rests on three findings:

• Laboratories are responsible for only 0.1–1.0% of the total hazardous waste that is generated in the United States, according to EPA estimates. Industrial production contributes the preponderance of hazardous waste to the U.S. total.

• Laboratory waste is characterized by substantially lower volumes and enormously greater chemical diversity than that from industrial operations.

• The direct and indirect effects on laboratories of some federal and state hazardous-waste regulations, which are designed primarily for industrial operations, are out of proportion to the fraction of hazardous waste that laboratories generate and to the overall hazard it poses.

The recommendations for simplifying the procedures for the handling and disposal of laboratory chemicals have two objectives:

• Increasing the productivity and efficiency of laboratories by reducing the extensive recordkeeping now required, and

• Providing laboratories with an array of practical, safe, and economically feasible ways to dispose of laboratory chemicals in an environmentally responsible manner.

We propose changes in regulations dealing with several areas of disposal of laboratory waste: (1) waste classification, (2) manifesting, (3) recordkeeping, (4) hazard reduction in the laboratory, (5) destruction of laboratory chemicals by incineration, and (6) interaction between federal and state regulations. In some of these areas laboratories face an array of EPA, DOT, state, and local regulations that can be inconsistent. These regulations are designed primarily for the management of large quantities of wastes, such as those generated in industrial operations. The establishment of a mutually consistent, interlocking regulatory approach among different agencies for handling and disposing of the small quantities of chemically diverse waste that are generated by many laboratories is a desirable goal.

WASTE CLASSIFICATION

The detail on individual chemicals currently required in recordkeeping and biennial reports seems unnecessary for laboratory waste. However, if reporting of the disposal of laboratory chemicals is required, we would propose that such chemicals be put into the seven classes listed below. These classes should be uniform for EPA, DOT, and state requirements for laboratory waste.

This classification should provide necessary and sufficient information for disposal of wastes on site, or for shipment off site either in lab packs for landfill or in combustible outer containers for incineration. The committee has used information developed by an American Chemical Society Task Force on RCRA in formulating this proposed classification. Under this proposal all laboratory wastes would be assigned to one of the following classes so that all manifests and recordkeeping could be based on this classification rather than on individual chemicals. The outer packs could contain more than one class, but the quantity of each class would be specified, and containers would bear the marking for each class as well as the required DOT marking. Containers with more than one class would be assigned an overall highest-hazard class, based on the hierarchy in which the individual classes are listed below. Mixing of classes within a single outer container would be limited to compatible chemicals, with the possible exception of class C (see proposed definition).

Classes of Laboratory Chemicals

A. Reactive	D. Ignitable
B. Toxic	E. Corrosive, acid
C. Miscellaneous Laboratory Chemicals	F. Corrosive, base
	G. Oxidizers

Class C, Miscellaneous Laboratory Chemicals, is proposed for disposal of chemicals that are found or generated in many laboratories in small individual quantities, such as residual samples from small-scale tests, minor by-products from reactions, residues from analytical procedures and chromatographic separations, filter paper, chemicals that may be contaminated with minor amounts of hazardous substances, and partially used small containers of reagents. Many of these types of materials are not well characterized, and the size of the sample is often so small that it precludes characterization. This committee believes that no useful purpose is served by listing what may be more than 100 individual small samples in a lab-pack-type container. With the restriction that reactive and oxidizing chemicals be excluded from such packs, there should be no significant hazard in landfilling or incinerating such packs. If a suitably small upper limit were to be put on the size of individual samples in such a pack, the prohibition on chemically incompatible chemicals in such a pack could be replaced by a prescription for the exercise of sound chemical judgment as to the types of chemicals being included in the pack. Since such packs would often contain at least some chemicals of unknown toxicity, they would usually be handled as containing toxic material.

MANIFESTING

This committee favors adoption of a uniform manifest that would obviate the need for additional or overlapping state manifests.

A simplified manifest for chemicals in a lab-pack-type pack could comprise the following:

A. A packing list of the quantities of chemicals in each class (see preceding section) in the pack.

B. The DOT proper shipping name for the pack.

C. Certification that the pack contains no individual container larger than 20 L (or 5 gallons); a lower maximum size would apply for Class C packs.

D. Certification that the pack contains no incompatible chemicals, with the possible exception of Class C packs.

E. For Class C packs:

(a) Certification that the pack has no containers of reactive or oxidizing chemicals.

(b) Certification that the pack has no individual containers larger than an established size limit.

(c) If the pack contains small samples of possibly incompatible chemicals, certification that sound chemical judgment was exercised in selecting the contents of the pack.

RECORDKEEPING

The recordkeeping requirements for generators of laboratory wastes who ship their wastes off site should comprise the following:

A. Retention of copies of packing slips for each individual pack for 3 years.

B. Retention of a list of the number of packs in each laboratory waste class shipped, with date of shipment, identification of carrier, and final destination. This list should satisfy reporting requirements.

For generators who dispose of wastes on site, the requirement should only be for recordkeeping of the quantities of wastes in each laboratory waste class disposed of and the method of disposal.

HAZARD REDUCTION

Chapter 6 provides procedures for the destruction/detoxification in the laboratory of small quantities of many types of hazardous chemicals, particularly those that cannot be put into a landfill because of the characteristic of reactivity. These procedures, some of which have been in common use for decades, are designed to convert hazardous chemicals into nonhazardous substances. Use of the laboratory procedures of Chapter 6 should be encouraged. Nothing should be done to change the common understanding among scientists that

• A laboratory procedure for the reduction or destruction of the hazard characteristic of a chemical is part of the experiment and does not constitute "treatment" in the regulatory sense, and
• Products resulting from the application of such a laboratory procedure to a listed chemical are not automatically considered to be hazardous unless they exhibit any of the characteristics of a hazardous waste.

DESTRUCTION OF LABORATORY CHEMICALS BY INCINERATION

The destruction of laboratory chemicals by incineration has the advantage over land disposal that the chemicals are destroyed with little potential for long-term harm to health and the environment if combustion emissions are properly controlled and the ash properly disposed of. However, many laboratories find that installation of an incinerator is not practical and that commercial incinerator operators will not accept their chemicals because of the small volume and chemical diversity. The laboratories are therefore constrained to use landfill disposal facilities.

Small-scale incinerators of various designs are commercially available, but their suitability for the disposal of hazardous waste has not been established or documented. Guidelines for the design and operation of small incinerators for hazardous waste are needed to assist laboratories in the selection of units and regulatory agencies in the granting of permits for such units. A suitable regulatory climate would encourage the development and use of such units, which could bring about a substantial reduction in the volume and numbers of laboratory chemicals that now go to landfill disposal. However, this committee believes that the necessary design and testing work for such development is not likely to occur if these units have to conform to the current standards for hazardous-waste incinerators, particularly the requirements for trial burns. Provision should be made for hazardous waste combustion units that

• Are designed to incinerate chemicals in quantities up to about 25 kg/hour.
• Are qualified to incinerate ignitable, corrosive, and listed hazardous waste.
• Are qualified to incinerate listed acute hazardous waste in quantities up to about 5% of the total fuel feed.
• Can be granted an operating permit on the basis of results of testing a prototype of a standard design or of meeting well-defined design and operating criteria.

FEDERAL AND STATE REGULATIONS

This committee is cognizant of the movement toward delegating to state environmental agencies the authority for enforcing hazardous-waste management regulations. Further movement in this direction could evolve into an array of 50 different sets of hazardous-waste management regulations. Basis for this view can be found in the report *State Integrated Toxics Management—Fact and Challenge*, March 1981, prepared by the National Governors Association for EPA, which illuminates the lack of uniformity among states in the area of hazardous-waste management. The committee believes that the delegation of authority to states should be accompanied by strong efforts to introduce reasonable uniformity and sound scientific principles into state regulations.

The Uniform Hazardous Waste Manifest proposed by EPA and DOT would be a useful step toward achieving uniformity in one area of hazardous-waste management. As another example, prospects for designing small units for incineration of laboratory quantities of hazardous chemicals are not likely to be realized unless individual states are willing to accept federal guidelines for the design, approval, and operation of such units.

REFERENCES

1. 45 FR, 10 May 1980. Copies of some of these regulations in effect as of 1 April 1983 are in Appendix A. Amendments to these regulations will be found in future issues of 40 CFR Subchapter I (see Appendix M).
2. NRC Committee on Hazardous Substances in the Laboratory, *Prudent Practices for Handling Hazardous Chemicals in Laboratories*, National Academy Press, Washington, D.C., 1981, pp. 5–6.

1 A Waste Management System for Laboratories

I. INTRODUCTION

Laboratory waste is generated in operations of widely different sizes and complexity, and the amounts and varieties of wastes generated vary accordingly. Waste management systems for a single-laboratory operation, for a small college, and for a large university or industrial complex must obviously have substantial differences in detail, but each system should have in common certain basic characteristics. Four elements that are essential to any laboratory waste management system are (1) commitment of the laboratory chief executive to the principles and practice of good waste management, (2) a waste management plan, (3) assigned responsibility for the waste management system, and (4) policies and practices directed to reducing the volume of waste generated in the laboratory.

II. EXECUTIVE COMMITMENT

The chief executive of a laboratory and his or her immediate subordinates must be openly committed to support of the waste management system. This commitment should include written support of the waste management plan, organizational responsibility for it, and willingness to allocate personnel and financial resources to the implementation of it. It is equally important that personnel at all levels of the laboratory organization—department heads, supervisors, academic faculty—exhibit a sincere and open interest in the waste management plan. This support must be continuing; it is not enough to support the plan at its outset and to assume that it will then operate. The success of the plan depends ultimately on the participation and cooperation of laboratory workers, which will be conditioned by their perception of management commitment. A program that is perceived as having only *pro forma* support will be ignored by laboratory personnel.

III. WASTE MANAGEMENT PLAN

It is the responsibility of laboratory management to see that a plan is developed for the handling of surplus and waste chemicals in the laboratory. The plan should be tailored to the operations of the laboratory, should be based on sound judgment, and should conform to the requirements of the U.S. Environmental Protection Agency Resource Conservation and Recovery Act (EPA RCRA) regulations as well as pertinent state regulations, which may be more restrictive. Written policies and procedures should be prepared to cover all phases of waste handling, from generation to ultimate safe and environmentally acceptable disposal. Although the necessary documents should be organized and written by the individuals who will be responsible for their implementation, there is much to be gained by having laboratory personnel participate in the planning process. Useful suggestions on waste handling may well come from those who are directly involved in working with chemicals. Moreover, their cooperation is essential to the successful working of the plan and is more likely to be forthcoming if they have had a part in formulating it.

The documents should describe all aspects of the system for the particular laboratory and should spell out responsibilities and specific procedures to be carried out by laboratory personnel, supervisors, management, and the waste management organization. An especially important document is a manual that summarizes those policies and procedures that directly affect laboratory personnel; copies of such a manual (which for a small laboratory may be relatively brief) should be available to all laboratory personnel. The plan should include provision for refresher training and auditing of all groups to assure continued compliance. The waste management plan should be reviewed at regular intervals to be sure that it conforms to current regulations and that it covers changes that may have occurred in the laboratory operations. The review should preferably be done every 6 months and in any event at intervals not greater than 2 years.

IV. ASSIGNED RESPONSIBILITY FOR THE WASTE MANAGEMENT SYSTEM

A successful waste management system requires a team effort among the following:

- Laboratory managers
- Laboratory supervisors
- Laboratory personnel
- Stockroom personnel
- The waste management organization
- The safety and health organization

The responsibilities of each of these groups should be set forth clearly in the waste management plan. Small laboratories may not have organizational subdivisions that correspond to each of these functions. However, the functions do exist, and the people who carry them out must assume the responsibilities assigned by the waste management plan.

Responsibility for implementing the waste management system must be specifically assigned, and the responsible individual should report directly to a high-level executive of the laboratory. The assignment of waste management responsibility to a subset of another part of the laboratory service organization could cause problems. The waste management organization can vary from one to many people, depending on the size and complexity of the laboratory. In some laboratories it may be necessary to make a part-time assignment; if so, care must be taken that the other duties of such an individual do not detract from waste management.

It is important that the individual in charge of waste management have enough knowledge of chemistry and of laboratory work to be able to understand the problems faced by laboratory personnel in their part of the system and to make chemically sound judgments about specific waste-handling situations. A waste manager who does not have a broad chemical background should have available a consultant who is chemically knowledgeable. The waste manager must be thoroughly familiar with and understand EPA regulations on hazardous waste and U.S. Department of Transportation (DOT) regulations on transportation of chemicals and must keep abreast of changes in them. In addition, state and local regulations also affect handling and disposal of waste, and the waste manager should establish and maintain contact with the responsible people in the local or regional offices of these regulatory organizations. The regulatory aspects of many waste-disposal problems can best be solved by working them out in advance with the agencies. In addition, waste-disposal contractors may impose specific requirements. The laboratory waste manager must be aware of the many levels of constraint that apply to the laboratory operation.

The waste management organization is responsible for setting up, maintaining, and inspecting waste accumulation sites, for disposal of waste, for providing advice on special disposal problems, and for training and refresher sessions for personnel.

Laboratory personnel, who are the generators of waste, must have general knowledge of the hazard characteristics of their waste and are in the best position to identify any of their chemicals that might pose unusual hazards. They are responsible for putting waste in the proper containers in accumulation sites. They must provide the information needed for proper maintenance of these sites and for recordkeeping purposes.

Laboratory supervisors must monitor laboratory operations and waste-accumulation sites to check on proper performance under the plan.

The responsibilities of stockroom personnel will depend on the size and complexity of the laboratory. Their activities may include putting dating labels on chemicals whose use is time-limited after opening, monitoring stocks of such chemicals, and operating an exchange clearinghouse for unneeded chemicals (see Chapter 4).

Even though these responsibilities in the operation of the waste management system are spelled out in the plan, they should be reinforced by training and refresher sessions. These should be set up and run by the waste management organization. Their frequency should be determined by the number and types of problems encountered in the operation of the system; by changes in regulations or in the nature of the laboratory operation; and, for new workers, by the rate of turnover of laboratory personnel.

V. REDUCTION OF VOLUME OF WASTE

Policies and practices directed toward reducing the volume of waste generated in the laboratory and toward avoiding special disposal problems should be integral parts of the waste management plan. Some examples are given in the following sections.

A. PLANNING EXPERIMENTS

The planning of every experiment should include consideration of the disposal of leftover starting materials and of the products and by-products that will be generated. Questions to be considered in such planning include the following:

- Can any material be recovered for reuse? (See Chapter 4.)
- Will the experiment generate any chemical that should be destroyed by a laboratory procedure? What procedure? (See Chapter 6.)
- Can any unusual disposal problem be anticipated? If so, inform the waste management organization beforehand.
- Are chemicals being acquired in only the quantities needed?
- Is there a possibility of replacing a hazardous reagent or solvent with one that is less hazardous?

B. REDUCTION OF THE SCALE OF EXPERIMENTS

The use of microtechnology in the study of chemical reactions can lead to significant savings in costs of chemicals, energy, apparatus, and space. In addition, such technology makes it possible to optimize on a small scale the conditions for a reaction that is to be carried out on a preparative scale so that the latter gives a high yield with minimal by-products. It is now technically feasible to run many reactions with much smaller quantities of chemicals than were needed 25 years ago. Some of the technical advances that have made this possible are the following:

1. Fast, microprocessor-based, top-loading balances that are sensitive to 0.1 mg;
2. Chromatographic techniques, such as high-performance liquid, gas, size exclusion, and ion exchange, that can cleanly separate and purify milligram quantities of substances;
3. Sensitive spectrometers that can analyze milligram, and sometimes microgram, quantities of substances;
4. Microscale glassware, including pipettes, burettes, syringes, reactors, and stills for handling reagents and reaction products;
5. Flow and transfer systems based on small internal-diameter metal and plastic tubing that make it possible to study flow-type reactions, catalysts, and multistep reactions on a very small scale, even under pressure.

C. CONTROL OF REAGENTS THAT CAN DETERIORATE

Indefinite and uncontrolled accumulation of excess reagents can create storage problems and safety hazards. These problems can be alleviated, and purchase costs saved, by instituting an excess-chemicals store to which laboratory workers can go for chemicals instead of ordering new material (see Chapter 4, Section IV). Chemicals that are prone to deteriorate with time can pose increasingly difficult disposal problems if allowed to accumulate. The waste management system (see Chapter 1) should provide a mechanism for periodic checking for such reagents.

Reagents that react readily with oxygen or water are prone to deteriorate when stored for long times after the original container has been opened. Deteriorated samples of water-reactive chemicals (see Appendix G) and pyrophoric chemicals (see Appendix H) should not be allowed to remain in the laboratory. Severe hazards can be created by peroxide-forming chemicals (see Appendix I) that have not been dated after opening the original container or that have exceeded the storage time limit after opening. Procedures for laboratory destruction of all three of these classes are given in Chapter 6.

D. MAINTENANCE OF REAGENT LABELING

Deterioration of labels is a common occurrence on old reagent containers. If the reagent or container has not deteriorated, the container should be relabeled if its identity is certain. Some suppliers of laboratory reagents are moving to add additional identification information, such as the CAS Registry Number, to reagent labels. This practice can be of help in identifying a reagent when the printed name has become uncertain, and this committee encourages the general adoption of this practice. However, if a reagent container label has disappeared or become illegible, and the nature of the chemical is unknown, the chemical should be discarded. The method for proper disposal can usually be determined by simple laboratory tests (see Chapter 6, Section V).

E. PREVENTION OF ORPHAN REACTION MIXTURES

Laboratory glassware containing reaction mixtures of unknown nature, and sometimes of unknown origin, can pose difficult disposal problems. Such materials occur frequently in research laboratories, particularly in those that have a high rate of personnel turnover. The disposal of an unlabeled orphan reaction mixture can sometimes be guided by knowledge of the chemistry that was being done by a departed laboratory worker. Simple laboratory tests (see Chapter 6, Section V) will often provide enough information to guide safe disposal of the material. The waste management system should provide a procedure designed to prevent the occurrence of such orphan wastes (see Chapter 1).

Laboratories should require that all reaction mixtures stored in laboratory glassware be labeled with the chemical composition, the date they were formed, the name of the laboratory worker responsible, and a notebook reference. This procedure can provide the information necessary to guide the disposal of the mixture if the responsible laboratory worker is no longer available. It should be recognized, however, that such a procedure must be policed and that it cannot guarantee that unlabeled mixtures will not be left behind by a departing worker.

Laboratories that encounter severe problems with unlabeled orphan reaction mixtures may institute a checkout procedure that requires departing laboratory workers to identify any reaction mixtures that they have not disposed of and to provide the information necessary for safe disposal of them. Academic institutions that have a high rate of turnover among graduate students may find it necessary to require a financial deposit from incoming graduate students, to be returned at the time of their departure after it has been determined that no orphan reaction mixtures have been left behind.

VI. THE WASTE MANAGEMENT SYSTEM

The following section is designed primarily for laboratories that are seeking guidance in setting up a waste management system. Laboratories with a system in place also may find it useful in improving elements of their existing system. An organization that is setting up a waste management system can profit from the experience of others. The American Chemical Society (ACS) Office of Federal Regulatory Programs, Department of Public Affairs, has reprints of talks on the management of hazardous waste that have been presented at ACS Regional meetings; it also operates a clearinghouse for information on how laboratories have dealt with certain waste-disposal problems. Academic institutions may be able to get advice on specific disposal problems from local industry.

The waste management system for any laboratory should be tailored to (1) the volume and variety of wastes generated, (2) the number and locations of generating sources within the laboratory, and (3) the options chosen for the disposal of wastes. These parameters should be determined as the system is being set up.

The first two of these parameters require a survey of the entire laboratory by the waste system manager. For small laboratories a brief survey that includes discussion with the individual responsible for each segment of the operation may suffice. For large, diverse operations it may be desirable to precede the survey with a simple questionnaire. It is almost inevitable that such a survey will uncover waste problems resulting from practices that should not be allowed to continue in the laboratory and that must be solved at the outset, for example, caches of overage and forgotten chemicals of unknown origin or identity (see Chapter 6, Section V). The system must arrange for the disposal of these and should be designed to prevent such accumulations in the future.

The survey should include all segments of the institution. Some laboratories will find units in the organization that are using chemicals and generating hazardous waste without awareness of any hazard. This situation is particularly likely in academic institutions, where chemicals are used routinely in many areas, such as the biology, geology, electrical engineering, art, and physics departments and the health service, by personnel who have little or no training in chemistry.

The survey should be designed to reveal the following:

1. The volume and types of nonhazardous waste being generated.

2. Total volume of hazardous waste generated per month. In some laboratories the waste volume may be broken down between wastes that are regulated by EPA and other types of wastes, such as radioactive materials. The former volume determines whether the laboratory qualifies for the EPA Small Quantity Generator Exemption (see Appendix A, 40 CFR 261.5). However, this quantity exemption is not recognized in some states (see Appendix B), and consultation with state hazardous-waste agencies is essential.

3. What chemical types of hazardous wastes are being generated, and in what locations. This information will determine many characteristics of the waste management system, such as procedures for classification, segregation, and collection (see Chapter 2) and the methods to be chosen for disposal of various types of hazardous wastes.

The waste management system manager must decide what types of wastes the system will handle. In some institutions the system handles only hazardous wastes; in others it takes care of all types of wastes generated by the institution. In either case the operation will incur costs that must be met, either by direct charges to individual laboratory units based on waste volume or by inclusion in general laboratory overhead. Either method can be made to work. The direct charge based on waste volume provides an incentive for laboratory units to recover, recycle, and reuse chemicals that would otherwise become waste. On the other hand, some laboratories find that the direct charge is an inducement to evade the system. Some laboratories have adopted the practice of adding a surcharge to the cost of purchasing chemicals, which is applied to cover waste-disposal costs. Whatever charge method is used, consideration should be given to charging laboratory units directly for disposal of an unusual waste that poses a special disposal problem outside the normal ambit of the system.

The structure of the waste management plan will depend heavily on the route selected for waste disposal. Large organizations may dispose of some of their waste on site or in their own facilities off site. These laboratories, as well as others that do not have their own waste-disposal facilities, may choose to do their own packing and manifesting of waste for transportation and disposal off site by a contractor. On the other hand, the problems of meeting the complex regulatory requirements for packing and off-site transportation can be reduced or eliminated by contracting for such services with a commercial waste-disposal firm. Some firms will bring in packing materials and labels, pack laboratory waste, prepare manifests, and arrange for transportation and disposal of the waste. A laboratory that contracts for such services should satisfy itself that the transportation and disposal are done responsibly and legally because the laboratory can incur legal and financial liabilities for improper handling and disposal of its waste even if done by someone else (see Chapter 8).

The waste-disposal plan should be summarized in a manual that is understandable by and available to all laboratory personnel. New laboratory workers should be made familiar with this manual by being given a copy at the time of their arrival or through an early indoctrination meeting. The content of this summary will vary according to the nature of the laboratory operation and the methods selected for waste disposal. Topics that should be considered for inclusion in it are outlined below. A laboratory that uses an outside contractor to perform some or all of its waste disposal will not need to include some of these topics and may need to include others to conform to the requirements of the contractor.

(a) A written endorsement of the plan by the chief executive officer of the institution.

(b) Introduction, setting out the purpose of the manual.

(c) A flow chart, along the lines of Figure 1.1, that lays out the waste management system for the laboratory.

(d) Overview of the waste management system.
  1. Objectives.
     a. Protection of the environment and of the health and safety of employees (or students) and of the neighboring community.
     b. Laboratory practices for reducing types and quantities of hazardous waste.
     c. Compliance with regulations—federal, state, and local.

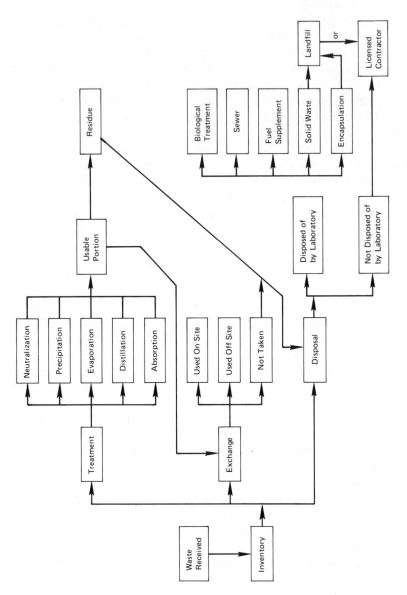

FIGURE 1.1 Example of a waste-disposal chart for a laboratory waste management organization.

2. Outline of types of wastes that are, or are not, handled by the waste management system.

3. Brief description of the responsibilities of laboratory personnel and of personnel in the stockroom, in the receiving room, and in the waste management organization (or those who carry out these functions).

4. Outline of routes of disposal for various types of wastes.

5. Planning for accumulation and disposal of wastes before they are generated.

(e) Procedures to guard employee health and safety in handling hazardous waste.

(f) Procedures for destruction of hazardous wastes in the laboratory.

(g) Classification, segregation, and collection procedures. The contents of this section will depend on the specifics of the system but might include the following:

1. A list of nonhazardous wastes generated by the laboratory (see Chapter 10, Section II).

2. Instructions for segregating incompatible chemicals (see Appendix E).

3. Instructions for segregating and collecting materials destined for the following:
   a. Exchange.
   b. Recovery and reuse.
   c. Use as a fuel supplement. (Materials destined for each of these end points should be listed, or illustrated, in appendixes to the manual.)

4. Instructions for packing and manifesting wastes for landfill disposal and incineration, if and to the extent applicable.

5. A description of the responsibility for and frequency of periodic inspection of waste-accumulation areas, including waste containers and safety/emergency equipment.

6. Instructions for disposing of wastes in the laboratory drain, including limits on quantities and chemical types (see Chapter 5 and Appendix K).

7. Instructions for handling wastes from teaching laboratories (if applicable).

8. Description of the interface between laboratory personnel and waste management personnel.
   a. Handling routine wastes.
   b. Handling wastes that require special disposal procedures.
   c. Handling spills and other accidents.

9. Review and training procedures.

    a. Frequency of review of the waste management system.

    b. Training and refresher sessions.

(h) Appendixes to the manual.

The manual can be made more readable by including some specific items in appendixes. Examples are:

    1. A glossary of terms and abbreviations.

    2. Lists of chemicals (see Appendixes E to I).

        a. Incompatible chemicals.

        b. Shock-sensitive chemicals.

        c. Water-reactive chemicals.

        d. Peroxide-forming chemicals, including time limits on retention after opening the original container.

        e. Monomers that could pose a hazard because of peroxide-induced, rapid polymerization.

    3. DOT classes (see Appendix D).

    4. (Optional) Some laboratories have developed computer lists of specific chemicals that are likely to appear as wastes in their operations and have included such a list as an appendix to their waste-system manual. Such computer lists should include CAS Registry Numbers, EPA Hazardous Waste Numbers, and DOT classes.

2 Identification, Classification, and Segregation of Laboratory Waste

I. INTRODUCTION

This chapter sets out guidelines that cover the identification, classification, and segregation of wastes generated in laboratories. These guidelines are based on the requirements of the U.S. Environmental Protection Agency Resource Conservation and Recovery Act (EPA RCRA) regulations. In some areas state and local regulations may impose additional requirements, and the waste-system manager should be sure that the laboratory system conforms to these as well. The classification and segregation procedures for the laboratory should be set forth in the laboratory waste management plan and summarized in the waste management manual (see Chapter 1).

II. IDENTIFICATION

The entire procedure of waste disposal starts with the laboratory worker, who could be, for example, a research chemist, an analyst, an engineer, a biologist, a geologist, or a laboratory teaching instructor. The laboratory worker must first decide if a material is no longer needed and is thus a candidate for disposal. Unneeded material does not become waste until a decision is made to discard it (see Appendix A, 40 CFR 261.2). All reasonable possibilities for use, recovery, recycling, or reuse of the material (see Chapter 4) should be considered. The laboratory plan should include guidelines on the types of chemicals that can be recovered, recycled, or reused by the laboratory.

Once material is declared waste, the first responsibility for guiding its proper disposal rests with laboratory workers. They are in the best position to know the degree of hazard posed by a material they have used or synthesized and must provide sufficient information on its characteristics to fit it into the correct channel for disposal. Accordingly, they should be familiar with the hazard characteristics by which wastes are classified and with the procedures used by the laboratory for segregating and collecting wastes. The laboratory waste management manual should provide enough information for this purpose.

It is first necessary to determine if a waste is a regulated hazardous waste. This identification is important in guiding the proper disposal of the material. In addition, the information is required to determine if the laboratory qualifies for a Small Quantity Exemption (see Appendix A, 40 CFR 261.5) and to assure that maximum storage time limits are not exceeded (see Appendix C). The EPA Small Quantity Exemption (see Appendix A, 40 CFR 261.5) limits for monthly generation and for storage of all regulated hazardous wastes are 1000 kg, cumulative total. The limits for acute hazardous wastes are lower. Some states have lower exemption limits for hazardous waste (see Appendix B).

III. CLASSIFICATION

The EPA regulations classify a waste as hazardous by (1) specifically including it on one of four lists and (2) defining four characteristics that the generator can use to determine whether a waste is hazardous. The four lists are (see Appendix A, 40 CFR 261 Subpart D) as follows:

1. *Hazardous Wastes from Nonspecific Sources*, which includes spent solvents that are common in laboratory operations and still bottoms from recovery of these solvents. Also included are sludges and solutions left over from several electroplating and metal-heat-treating operations, most of which contain cyanide.

2. *Hazardous Wastes from Specific Sources*, which includes primarily industrial wastes not usually associated with laboratory operations.

3. *Discarded Commercial Chemical Products, Off-Specification Chemicals, Containers, and Spill Residues Thereof.*

(a) *Acute Hazardous Wastes*, which are chemicals that are considered to represent an acute hazard. The EPA Small Quantity Exemption (see Appendix A, 40 CFR 261.5) limits for monthly generation and for storage of materials on this list are 1 kg, cumulative total. There are stringent requirements for disposal of used containers.

(b) *Hazardous Wastes*, which are chemicals that are considered to pose less hazard than acute hazardous wastes. Requirements for disposal of used containers are less stringent than those for used containers of acute hazardous chemicals.

If a waste is not listed, it is still a hazardous waste if it meets any of the following four characteristics; these characteristics can be determined by specific tests cited in the regulations (see Appendix A, 40 CFR 261 Subpart C). Alternatively, if knowledge of the source or properties of a material indicates that it may have any of these characteristics, the material can be declared hazardous without being tested.

1. *Ignitability*. The definition of this characteristic includes the following:

(a) Liquids, other than aqueous solutions containing less than 24% alcohol by volume, that have a flash point below 60°C.

(b) Materials other than liquids that are capable, under standard temperature and pressure, of causing fire by friction, absorption of moisture, or spontaneous chemical changes and, when ignited, burn so vigorously and persistently as to create a hazard.

(c) Flammable compressed gases as defined by U.S. Department of Transportation (DOT) regulations 49 CFR 173.300. This definition includes gases that form flammable mixtures at a concentration of 13% or less in air or that have a flammable range in air wider than 12% regardless of the lower limit. The definition includes other characteristics that must be determined by specified tests.

(d) Oxidizers as defined by DOT regulations: a substance, such as a chlorate, permanganate, inorganic peroxide, or nitrate, that readily yields oxygen to stimulate the combustion of organic matter (see DOT regulation 49 CFR 173.151).

2. *Corrosivity*. Aqueous solutions that have a pH equal to or less than 2 or equal to or greater than 12.5 are corrosive. Other liquids are defined as corrosive if they corrode SAE 1020 steel at a rate greater than 6.35 mm/year at 55°C.[1] A characteristic for corrosive solids is not specified in the regulations, and therefore a solid need not be classified as an EPA-regulated corrosive waste.

3. *Reactivity*. This classification includes substances that react with water violently or to produce toxic gases or explosive mixtures, substances that are unstable, explosives, and substances that contain cyanide or sulfide that generate toxic gases when exposed to a pH in the range 2–12.5.

4. *EP (Extraction Procedure) Toxicity.* This characteristic is defined by a prescribed test procedure for water extraction of the waste at a pH $\leq$ 5, controlled by addition of acetic acid. The extract is analyzed for concentrations of eight elements or ions [As, Ba, Cd, Cr(VI), Pb, Hg, Se, Ag] and six chlorinated organic substances (see Appendix A, 40 CFR 261 Subpart C). This characteristic cannot apply to a waste that is known not to contain any of these 14 materials.

To determine if a waste is regulated as hazardous it is necessary to check all applicable state regulations as well as those of the EPA. Authority to administer hazardous-waste disposal has been delegated to some states by EPA, and many of these have additional listed wastes and additional characteristics.

The EPA regulations place on the waste generator the burden of determining whether a waste is hazardous and in what hazard classification it falls. Testing is not required specifically by the regulations, and in most cases the laboratory worker should be able to provide enough information about the waste to allow the hazard classification to be assigned. If the waste is not a common chemical with known characteristics, enough information about it must be supplied to satisfy the regulatory requirements and to be certain that it can be handled and disposed of safely. For many wastes only the principal components need be specified; however, if the waste contains a carcinogen or a heavy metal, this information should be supplied. The information needed to characterize a waste also depends on the method of ultimate disposal.

IV. SEGREGATION

A. Labeling

Classes of waste must be properly segregated for temporary accumulation and storage as well as for transportation and disposal. Accordingly, all wastes must be properly labeled before being moved into the disposal sequence. The label should contain sufficient information to assure safe handling and disposal, including the initial date of accumulation and chemical names of the principal components and of any minor components that could be hazardous. The label should also indicate whether the waste is toxic, reactive, corrosive to skin or metal, flammable, an inhalation hazard, or a lachrymator. An example of a two-sided label on which laboratory personnel can provide necessary information to the waste management organization is shown in Figure 2.1. Some laboratories use labels that provide information on cleaning up accidental spills; an example is shown in Figure 2.2.

REQUEST FOR DISPOSAL OF MATERIAL

ORIGINATOR (PRINT NAME)		DATE
LOCATION	ROOM NO.	PHONE NO.

This material is properly described, has descriptive labels, and is in a proper container for handling by Purchasing.

_____
PROFESSIONAL'S SIGNATURE

Identification of Material
Use Generic Name — Do Not Abbreviate or use Chemical Formula

COMPONENT (List ALL Compounds)	% or ppm
1.	
2.	
3.	
4.	
5.	
6.	
7.	
8.	

Total Quantity   Control No.

(PURCHASING USE ONLY)

Hazard Classification

☐ **Extremely Hazardous** *(Check hazards below)*
☐ **Hazardous** *(Check hazards below)*
☐ **None of the Hazards Listed Below Apply**
☐ **Can be Blended to Fuel**

HAZARDS *(Defined on Back)*
Check all Hazards that Apply

Explosive . . . . ☐
Compressed Gas – Flammable . . . . ☐
Compressed Gas – Nonflammable . . . . ☐
Poisonous Gas or Liquid - Class A . . . . ☐
Ignitable Liquid - Est. Flash Point (CC): _____°F . . . . . . ☐
Organic Peroxide . . . . ☐
Flammable Solid . . . . ☐
Corrosive - Est. pH _____ . . . . ☐
Poisonous Liquid or Solid - Class B . . . . ☐
EP Toxic *(List all components shown in definition)* . . . . ☐
Oxidizer . . . . ☐
Reactive . . . . ☐
Irritant - Intense Lacrymator . . . . ☐
Irritant - Skin Irritant/Sensitizer . . . . ☐
Other Hazard *(Describe Below)* . . . . ☐

_____
_____
_____
_____

HAZARDOUS MATERIALS DEFINITIONS AND EXAMPLES

Extremely Hazardous Materials — A Shell Development Company designation for materials whose disposal requires *special consideration in handling, labeling and/or packaging because of particular hazards.* Included herein are the materials classified as explosive, poisonous gases class A, flammable liquids and highly flammable solids, the most poisonous class B liquids or solids, potent carcinogens, potent neurotoxins and very irritating materials.

Candidate Fuel — Stable liquid hydrocarbons or oxygenates, such as alcohols and ketones. These materials must *not* contain *halogenated* compounds; sulfur content must be less than 0.3%w. (Examples: Used lubricating oil, slopped fuel, solvents and solvent mixtures, etc.)

Explosive — Any chemical compound, mixture, or device which could undergo a substantially instantaneous release of gas or heat by detonation or rapid combustion. Disposal of explosives requires special procedures. Call HS&E, WRC Ext. 7073 or BRC Ext. 2334, for further information. (Examples: commerical explosives, polynitro-organic compounds, some azides, and diazo compounds.)

Compressed Gas - Flammable — Any flammable gas having in the container an absolute pressure exceeding 40 psi at 70°F or 104 psi at 130°F; or any flammable liquid having a vapor pressure > 40 psi at 100°F. (Examples: hydrogen, methane, propylene, hydrogen sulfide, butane, vinyl chloride, carbon monoxide.)

Compressed Gas - Nonflammable — Any nonflammable compressed gas having in the container a pressure exceeding 40 psi at 70°F or 104 psi at 130°F. (Examples: nitrogen, argon, oxygen, certain freons, carbon dioxide.)

Poisonous Gas or Liquid - Class A — Poisonous gases or liquids so dangerous that a very small amount of gas, vapor, or liquid mixed with air is dangerous to life. (Examples: cyanogen, hydrogen cyanide, nitric oxide, phosphine, arsine, hydrogen selenide.)

Ignitable Liquid — Any liquid having a closed cup flash point < 200°F (Includes flammable [FP < 100°F] and combustible [FP 100-200°F]. (Examples: FP < 20°F; ethylene oxide, acrolein, allyl chloride, acetaldehyde, methyl mercaptan, carbon disulfide, diethyl ether, n-pentane, n-hexane, benzene, acetone. FP 20°-100°F; toluene, xylenes, n-heptane, n-nonane, methyl alcohol, pyridine. FP 100°-200°F; n-C10 through n-C13 paraffins, diethylbenzenes.) *Flash points must be shown!*

Organic Peroxide — Organic compounds containing the -O-O- group. (Examples: acetyl and benzoyl hydroperoxides, di-t-butyl peroxide, alkyl and aryl hydroperoxides.

Flammable Solid — Any solid material other than an explosive, which is liable to cause fire through friction, adsorption of moisture, spontaneous chemical changes, which is pyrophoric or which can be ignited readily and when ignited burns so vigorously and persistently as to create a serious hazard. (Examples: metal hydrides, alkali metals, metal alkyls, metal carbonyls, finely divided metals, Raney nickel, low MW phosphines, elemental phosphorus pentasulfide or pentachloride, calcium metal, sodium hydrosulfite, activated charcoal.) Some of these flammable solids are also *extremely hazardous or reactive* and they should be so labeled.

Corrosive Material — Any liquid or solid that causes irreversible destruction of human skin tissue, or a liquid that has a severe corrosion rate on steel. (Examples: strong mineral acids, liquid aliphatic acids, inorganic and organic acid chlorides, halo-acetic acids, benzyl bromide and chloride, antimony pentachloride, phosphorus trichloride, alkali metal hydroxides, titanium tetrachloride, trimethylchlorosilane, bromine.) *pH must be shown!*

Poisonous Liquid or Solid - Class B — Materials less poisonous than Class A poisons yet known to be so toxic to man as to afford a hazard to health by inhalation, skin absorption or ingestion. Materials which in the absence of adequate data on human toxicity are presumed to be toxic based on results with test animals. (Examples: acrolein, allyl alcohol, acrylonitrile, aniline, toluene-diisocyanate, strong bases and acids, phenols, beryllium compounds, alkaloids, metal cyanides, phosgene, nitrobenzene.)

EP Toxic — Materials from which the following contaminants, even at very low concentration (ppm/ppb), may be extracted by an aqueous acetic solution: arsenic, barium, cadmium, chromium, lead, mercury, selenium, silver, endrin, lindane, methoxychlor, toxaphene, herbicide 2,4-D, herbicide 2,4,5-T. (Examples: dichromate cleaning solution, catalysts containing any of the listed metals at greater than ppm concentration, discarded metal salts.)

Oxidizer — A substance that yields oxygen readily to stimulate the combustion of organic matter. (Examples: inorganic nitrates, perchlorates, permanganates, chromates and hypochlorite salts, concentrated organic and inorganic peroxides, chromium trioxide.)

Reactive — Explosive, unstable, reacts with water violently or forms explosive mixtures or generates dangerous quantities of toxic gases such as hydrogen sulfide and hydrogen cyanide with water at any pH. (Examples: cyanide or sulfide containing material, commercial explosives, organic peroxides, titanium tetrachloride, alkali metals, etc.)

Irritant
a) Strong lachrymators — direct or as the result of decomposition (Examples: Chloroacetophenone, bromobenzyl cyanide).
b) Strong skin sensitizing agents. (Example: Urushiol and other related alkylated polyphenols, epichlorohydrin and certain resins containing epichlorohydrin, cantharidin.)

Other Hazards — Includes hazards not described above. (Examples: chronic toxins, embryo-fetotoxins, etiologic agents (i.e., a viable micro-organism or its toxin which causes human disease.)

FIGURE 2.1

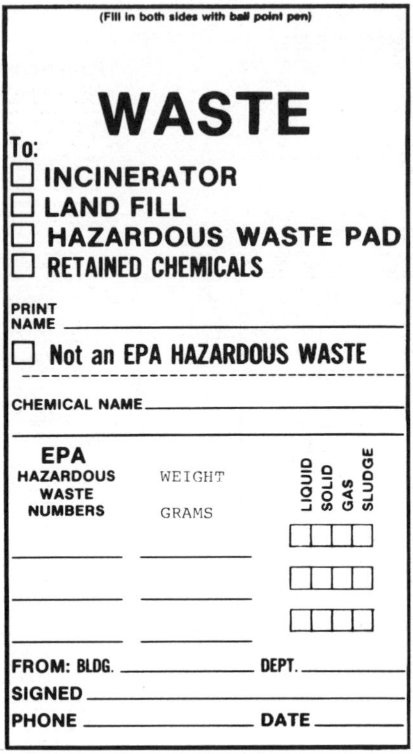

```
(Fill in both sides with ball point pen)

WASTE
To:
☐ INCINERATOR
☐ LAND FILL
☐ HAZARDOUS WASTE PAD
☐ RETAINED CHEMICALS

PRINT
NAME _____

☐  Not an EPA HAZARDOUS WASTE
- - - - - - - - - - - - - - - - - - - - - - - - - - - -

CHEMICAL NAME_____

     EPA
  HAZARDOUS     WEIGHT        LIQUID
    WASTE                     SOLID
   NUMBERS      GRAMS         GAS
                              SLUDGE
                              ☐☐☐☐
 _____    _____
                              ☐☐☐☐
 _____    _____
                              ☐☐☐☐
 _____    _____

FROM: BLDG. _____ DEPT. _____
SIGNED _____
PHONE _____ DATE _____
```

FIGURE 2.2

Some laboratories, particularly research laboratories, use and syn-thesize many unusual chemicals that can become waste. In general, chemists involved with such chemicals should know qualitatively whether they are ignitable, corrosive, or reactive. If large quantities of a chemical waste for which these hazard characteristics are not known are being generated, analytical tests can be run to determine them. However, for typically small quantities of laboratory chemicals, formal analysis is not warranted. Laboratory samples of chemicals with unknown hazard char-acteristics, including orphan wastes (those generated by workers who have departed from the laboratory, leaving unidentified materials be-hind), can be tested on a small scale in the laboratory for flammability with a flash tester, pH if aqueous by using pH test paper, reactivity with water or air, and peroxides or other oxidizing property with potassium iodide (see Chapter 6, Section II.P). The one hazard characteristic that cannot be readily determined is toxicity, although the probability that a chemical is or is not toxic can sometimes be inferred by analogy to closely related chemical structures. In the absence of knowledge or a basis for judgment a waste should be assumed to be toxic and labeled accordingly.

A laboratory that does not qualify for the Small Quantity Generator Exemption (see Appendix A, 40 CFR 261.5) and that disposes of EPA-regulated hazardous waste on site must have a waste-analysis plan that describes procedures used to obtain information required for safe handling and disposal of wastes. Although the regulations do not require an analysis of waste disposed of off site, the practical fact is that waste-handling firms require some definitive information on the composition or characteristics of wastes they handle. The type and amount of information required depend on the mode of disposal; this information will usually accompany the shipping manifest. Chemical analysis of a waste is more practical and important for a large-volume industrial waste that is being generated on a regular basis than it is for the small volumes of the chemically diverse wastes generated in the laboratory. For the latter it often suffices to characterize without an analysis but rather from knowledge of what went into it, e.g., "hydrocarbon mixture," "flammable laboratory solvents," "chlorobenzene still bottoms." The professional expertise, common sense, judgment, and safety awareness of trained professionals performing chemical operations in the laboratory most often put them in a position to judge the type and degree of hazard of a chemical. A trained professional is an individual qualified by training and experience to evaluate the hazard posed by a chemical and to direct its disposal so as to protect (within the limits of scientific knowledge) against such hazard.

B. Disposal of Wastes in Sewer Systems

Limited quantities of some wastes can be disposed of in *sanitary* sewer systems, which will destroy the wastes, but never in *storm* sewer systems. A sanitary sewer is one that is connected directly to a water-treatment plant, whereas a storm sewer usually discharges eventually into a stream, river, or lake. Guidelines for types and quantities of chemicals that can be disposed of in a sanitary sewer are given in Chapter 5.

C. Accumulating Waste for Disposal

The first step in the disposal sequence usually involves accumulation or temporary storage of waste in or near the laboratory. The requirements and limits for storage of wastes under the EPA regulations are outlined in Chapter 3. Although the following practices should meet regulatory requirements, specific state and local regulations pertinent to the laboratory site should be checked. An EPA Hazardous Waste Number (see Appendix A, 40 CFR 261.12) is required for off-site transportation and disposal of all regulated hazardous wastes.

Except when a single chemical is being accumulated for recovery, waste accumulation generally involves several chemicals in a container. ONLY COMPATIBLE CHEMICALS SHOULD BE PUT IN ANY CONTAINER, WHETHER PACKAGED SEPARATELY OR MIXED. In this context, "compatible" connotes the absence of potential for chemical interaction. Guidelines on INCOMPATIBLE CHEMI-CALS are given in Appendix E. The two common practices for accumulation of wastes are (1) mixing compatible chemicals in a waste container and (2) accumulating small containers of compatible waste in a larger outer container, e.g., a lab pack (see next paragraph). The method chosen and the scheme for segregating the wastes depend primarily on the intended mode of disposal.

If laboratory wastes are to be landfilled, the most common method of packaging is the lab pack (see Appendix A, 40 CFR 265.316). The procedure for preparing a lab pack is described in Chapter 10, Section V.B, and the regulatory requirements for manifesting and shipping lab packs are given in Chapter 8. The waste management plan and manual should include specific directions for preparing lab packs as well as the assigned responsibility for preparing them. On the other hand, if a contract disposal service that prepares the lab packs is used, the manual should give directions for segregating and labeling wastes in accordance with the contractor's requirements. The principal consideration in segregation of chemicals for landfill disposal is compatibility. The laboratory waste management manual should contain guidelines for, or a list of, incompatible chemicals (cf. Appendix E). In addition, chemicals that have the hazard characteristic of reactivity (except for cyanide- and sulfide-bearing reactive wastes) and explosives cannot be put into a landfill (see Appendix A, 40 CFR 265.316).

The method chosen for segregation and accumulation of wastes destined for incineration depends on the design of the incinerator and its waste feed mechanism, which vary widely. Some incinerators can handle only bulk liquid wastes, whereas others accept solid or packaged wastes such as fiber packs, glass and plastic bottles, and some even steel cans or drums. A more detailed discussion of incineration capability is found in Chapter 9.

Incinerators that accept only liquid wastes either blend them with other fuels or incinerate directly. In either case the disposer generally prefers to pump the contents from the container, and small containers are less desirable than the standard 55-gallon drum. Incinerators that accept solid waste generally prefer to incinerate without removing the waste from the container, avoiding the hazards of opening containers and handling wastes. Some facilities will accept a wide variety of containers, including individual bottles. Others prefer to accept wastes in fiber packs, which is a combustible version of the lab pack that is used in landfills. Many incinerators or their belt-feed conveyors are loaded by hand, and fiber packs in the size range 5–10 gallons (20–40 L) are preferred.

Since incinerators must be operated within specific ranges of fuel heat content, the approximate heat of combustion of mixed liquid wastes should be known so that appropriate feed rates can be determined. The approximate heat of combustion of fiber packs also should be known and should not exceed a specific limit because these packs are fed as units. The heat of combustion of wastes can be estimated from their constitution and quantity, using values in any standard handbook. Incinerator operators can provide values to be used for the heat of combustion of fiber packs and associated inert fillers. Explosive compounds should not be put into fiber packs for incineration. Some operators will incinerate some types of explosives by adding them to the incinerator feed in small quantities, preferably diluted with a flammable solvent or sawdust. Containers with more than about 100 mL of carbon disulfide cause an explosive hazard in incinerators and should be avoided. Procedures for disposal of laboratory quantities of carbon disulfide are given in Chapter 6, Section II.G.

If separate incinerator facilities for hazardous and nonhazardous waste are accessible, segregation of hazardous waste can be both cost-effective and environmentally prudent. Compared with, for example, incinerators for municipal waste, incinerators for hazardous waste require more expensive design, more careful operation, high costs for obtaining permits, and regulated treatment of ash and scrubber water. Consequently, incineration of nonhazardous waste in hazardous-waste facilities is costly.

Proper sorting of wastes destined for incineration is essential. Halogenated wastes should be kept separate from nonhalogenated wastes because the former must be burned in an incinerator with a scrubber that reduces emissions of halogens and hydrogen halides. More than trace amounts of brominated compounds also should be kept separately; they usually require special disposal procedures because bromine and hydrogen bromide are not removed effectively by most scrubbers. Some incinerators can handle brominated compounds at a controlled feed rate consistent with safe stack emissions; the alternative is to landfill these compounds in a lab pack. A typical segregation scheme for incineration involves separate fiber packs for halogenated organics, for nitrogen-containing compounds, and for other organics. If inorganics or other compounds that pose problems for the incineration facility are involved, they should be put in yet another fiber pack. Generally the disposer will specify acceptable ranges for size and heat of combustion of fiber packs as well as the type and amount of information required on their contents.

Accumulation of laboratory wastes for disposal must be carried out in accordance with the written waste management plan, which should include all safety factors (chemical incompatibilities are particularly important), regulatory requirements, and factors specific to the disposer and disposal method. A training program should be initiated and maintained to assure that all laboratory personnel understand the elements of the plan, such as accumulation procedures, safety procedures, and recordkeeping.

In general, the design of a waste accumulation point should follow the principles given for a waste storage area (see Chapter 3). It is especially important that the design conform to local fire codes. However, points for accumulation of small quantities of waste over a period of a few days need not be as elaborate as larger waste storage areas. In some laboratories—for example, teaching laboratories—modest quantities of single-type wastes can be accumulated temporarily in containers kept in a hood. Such containers should be clearly labeled for composition of their contents and for date. Large quantities of flammable wastes should never be accumulated in a closed, unventilated room; a vapor explosion in such a room will not be contained, and the room may become a bomb. An appropriate container for each category of waste in the waste accumulation plan should be available at all times. The container must be clearly marked to indicate the type of waste it can contain and the hazard associated with this category. It should be dated to indicate when accumulation started.

The accumulation area should be inspected regularly. The frequency of inspection can depend on the level of activity and the degree of hazard but generally should be at least weekly. The inspection should include the following:

1. Adherence to the accumulation plan;
2. Condition of containers;
3. Availability of containers;
4. Container dates, for both safety and regulatory requirements;
5. Adequate records of container contents;

6. Operation of safety equipment.

Documentation should be kept on the contents of each accumulation container unless the container is labeled to receive, for example, only a single type of waste. The specific data required will depend on the size and complexity of the laboratory operation. Where large quantities of different wastes are being accumulated, the records should include the amounts of each waste, the date each was placed in the container, the chemical identity or type, and any information that will help in assigning the hazard class. A waste identification (or analysis) form is helpful in assuring that the necessary information is included.

Although the same principles apply to smaller laboratories with simpler or fewer wastes, such laboratories may be able to work with simpler rules and requirements. For example, in a teaching laboratory where a specific experiment is being carried out by students, it should suffice to collect the wastes in containers labeled for each waste and to record the

HAZARDOUS WASTE

FEDERAL LAW PROHIBITS IMPROPER DISPOSAL

IF FOUND, CONTACT THE NEAREST POLICE, OR
PUBLIC SAFETY AUTHORITY, OR THE
U.S. ENVIRONMENTAL PROTECTION AGENCY

PROPER D.O.T.
SHIPPING NAME _____ UN OR NA# _____

GENERATOR INFORMATION:

NAME _____

ADDRESS _____

CITY _____ STATE _____ ZIP_____

EPA   EPA
ID NO. _____   WASTE NO. _____

ACCUMULATION   MANIFEST
START DATE _____   DOCUMENT NO. _____

HANDLE WITH CARE!
CONTAINS HAZARDOUS OR TOXIC WASTES
STYLE WM-6

FIGURE 2.3

total quantity and date. A similar system can be used where specific waste solvents are being accumulated for recovery. If the wastes are to go into a lab pack for landfill disposal, or into an analogous fiber pack for incineration, the accumulation record should include the information that is on the individual waste containers.

The waste management plan should cover the specific requirements and accumulation procedures of the laboratory. It is important that laboratory workers be provided with appropriate forms and codes to make it as simple as possible for them to comply with the requirements.

Before the waste accumulation container is transported either to the disposer or to another accumulation point, it must be labeled in accordance with DOT, EPA, and any necessary state shipping requirements. A sample label that contains the minimum information required for an EPA-regulated waste is shown in Figure 2.3. Requirements for off-site transportation of hazardous wastes are given in Chapter 8.

REFERENCE

1. EPA Manual of Test Procedures, method 1110; adapted from National Association of Corrosion Engineers Standard TM-01-69 (1972 revision).

3 Storage of Laboratory Waste

I. INTRODUCTION

Containers of waste chemicals collected from individual laboratories must often be placed in a temporary storage facility before being treated, disposed of, or stored elsewhere. This chapter deals with provisions that are required for storage of U.S. Environmental Protection Agency (EPA)-regulated waste. The same provisions should also be observed for storage of hazardous waste that is not EPA regulated: waste that does not meet any of the four EPA hazard characteristics (see Appendix A, 40 CFR 261 Subpart C) and that is not included on the EPA lists (see Appendix A, 40 CFR 261 Subpart D) but that is known to be or could be toxic. Temporary storage times should be kept as short as possible. The storage facility should be designed for total containment with as little as possible release to the environment.

Temporary storage of laboratory waste may be subject to Resource Conservation and Recovery Act (RCRA) requirements, and in some cases a permit may be required. The RCRA regulatory requirements are summarized in Appendix C, together with copies of the pertinent sections of the regulations in effect in 1982. EPA has published (3 January 1983) a proposed amendment to the storage regulations (see Appendix A, 48 FR 121, 3 January 1983) (the "Satellite Storage" amendment) that would allow temporary accumulation of up to 55 gallons (208

L) of hazardous waste [but not acute hazardous waste, see Appendix A, 40 CFR 261.33(e)] at or near a point of generation, without being subject to regulation. Certain restrictions apply (see Appendix A). The design of the storage facility must comply with all local regulations, including fire codes. The following sections outline prudent practices for storage of laboratory waste, regardless of whether it is regulated by the EPA. There are additional requirements both on facilities and on procedures to obtain a permit for storage of EPA-regulated waste for more than 90 days (see Appendix C).

II. PROTECTION FROM THE ELEMENTS

Waste containers should always be stored in an area or facility where the containers are protected from adverse weather. Extremes in heat can cause pressure buildup in containers with volatile liquid contents. Alternate heating and cooling may cause containers to breathe if they are not tightly sealed. If water is left standing on the container, it can be drawn into the container during a cooling cycle, resulting in container corrosion or internal chemical reactions. Adverse weather conditions can also cause deterioration of labels, tags, or other markings, which can create a hazard and could necessitate reanalysis of container contents.

A properly designed storage facility should have a roof to protect waste from sun and precipitation. Warning signs should be posted and walls or fences erected to protect against unknowing or unauthorized entry. The facility must, however, have proper ventilation and must provide access for fire or other emergency equipment. The facility should be located away from areas of high work density but close enough to be useful and to allow proper surveillance and security.

III. PRIMARY CONTAINMENT

Waste should be put into containers that are adequate to contain the waste through the storage period and until final disposition. If the waste is being stored prior to transportation, the use of appropriate U.S. Department of Transportation (DOT)-approved containers is required. Even if the waste is not to be transported, the DOT regulations provide excellent guidance for container selection (see Appendix D, Section II).

IV. SECONDARY CONTAINMENT

It is prudent to use secondary containment under storage areas to catch leaks, spills, and precipitation that becomes contaminated. These provisions are required in order for the storage area to obtain an EPA storage permit (see Appendix C). The storage area base must be impervious to the waste being stored and to rainfall. The base should be designed such that liquid from leaks or precipitation does not come in contact with the containers or at least so that such contact is minimal. This can be accomplished by sloping the base toward a sump or other collecting basin or by elevating the containers above the base. The containment should have levees, and the total capacity of the containment should be at least 10% of the total volume stored or equal to that of the largest container, whichever is greater.

Any material collected in a secondary containment area must be identified before being removed for treatment or disposal. Sometimes this identification is obvious, as in the case of rupture of a labeled container; in other cases it may be necessary to perform a chemical analysis prior to prescribing proper disposition of the material.

V. FIRE SUPPRESSION

The fire protection required for a storage area is often dictated by laboratory policies or by local codes and regulations. A storage area in which flammable materials are stored should have fire emergency equipment, which should include, at the least, an appropriate fire extinguisher, fire blankets, and breathing equipment. If large volumes of flammable materials are stored or if the storage is near areas where ignitable materials or structures are located, a sprinkler system is advisable.

VI. INSPECTION

An area used for storage of containers of EPA-regulated waste must be inspected at least weekly (see Appendix C). Even if a waste storage area does not contain EPA-regulated waste, periodic inspection is recommended. The inspection should follow a written plan and should include general housekeeping, all containers, and safety and emergency equipment. Containers should be inspected for leaks, corrosion, proper closure, labels, and segregation. Safety and emergency equipment should be inspected for general condition, expiration of reservice dates, and actual operation if possible. Secondary containment should also be checked and emptied if necessary. Any problems noted in an inspection must be reported promptly to the person responsible for the area and corrected immediately.

Inspections are more effective if a checklist is used and if the inspection is a dedicated operation. It is not prudent to perform inspections while putting waste into or taking waste out of the area. It is a particularly poor practice to sign off on an inspection on the basis of frequent visits to the area for other purposes.

VII. SEGREGATION OF WASTES

Segregation of incompatible materials (see Appendix E) in a storage area is essential. It is particularly important to segregate ignitables from oxidizers or sources of ignition. Other types of incompatibles should never be put in the same container; segregation of their containers is not always required, although it is desirable. Additional segregation requirements are often dictated by the ultimate disposal option (see Chapter 2).

VIII. RECORDKEEPING

The recordkeeping requirements for transportation, treatment, and disposal are extensive and depend on the ultimate fate of the waste. Requirements frequently involve a complete characterization of the waste, i.e., a chemical analysis of mixed wastes or a complete contents list for lab packs. While it may be tempting to generate the required information when the waste is moved out of storage, it is better practice to keep a running inventory of wastes and contents of waste containers in storage. This helps with the necessary paperwork, usually assures more accurate information, and is safer. It is not good practice to store containers whose contents are not known. A good way to collect the information is to establish a log-in/log-out procedure for the storage area.

IX. TRAINING

Inadequate training of personnel who use a waste storage facility is a potential source of trouble. Everyone involved should be thoroughly familiar with proper waste segregation, emergency equipment, emergency procedures, required inspections, and waste inventory procedures. A written plan and documented training sessions are required for facilities that store EPA-regulated waste (see Appendix C) and are a good idea for all facilities.

X. CONTINGENCY

A storage area for hazardous waste must have an emergency plan that covers foreseeable types of contingencies. Such a plan is required for facilities that store EPA-regulated waste (see Appendix C). The plan should be reviewed periodically, and drills should be conducted. The plan should identify responsible persons who can coordinate any emergency response. The plan should (*must* for RCRA-regulated facilities) be submitted to local police departments, fire departments, hospitals, and emergency response teams that could be called on to provide emergency services.

XI. CLOSURE

Any facility must be properly cleaned up and decontaminated, if necessary, after it is taken out of active service. This includes disposal of any remaining waste, cleanup of secondary containment, and disposal of any contaminated equipment or releases. If the facility has an EPA permit, the closure plan must be in writing, must be approved prior to closure, and must be certified as complete after closure.

4

Recovery, Recycling, and Reuse of Laboratory Chemicals

I. INTRODUCTION

One intent of the Resource Conservation and Recovery Act (RCRA), which provides the legislative basis for the U.S. Environmental Protection Agency (EPA) regulations on management of hazardous waste, is to encourage the recovery, recycling, or reuse of materials that would otherwise become wastes. This intent is recognized in the EPA definition of a solid waste as a material that is being discarded or being accumulated; stored; or physically, chemically, or biologically treated prior to being discarded. Use, reuse, reclamation, and recycling are explicitly excluded from the meaning of "discarded." To the extent that chemicals can be recovered, recycled, or reused safely at net costs less than the costs of disposal as waste, there is obvious economic incentive to do so. In addition, materials that are recovered, recycled, or reused do not become a problem in the environment.

Even though recovery, recycling, and reuse of laboratory chemicals may not currently be an economical option for a given laboratory, it is an option that should be continually assessed. It appears probable that disposal of hazardous waste in secure landfills will become more expensive and that the capacity of such landfill facilities may not keep up with demand. The prospects for easier access to incineration facilities that will accept the small volumes of chemically diverse laboratory wastes

44

are questionable. These factors may change the relative economics of recovery, recycling, and reuse. Recovery of laboratory chemicals in house can be a valuable educational experience in academic laboratories, when carried out under proper supervision.

II. RECOVERY OF VALUABLE METALS

The metal content of a number of materials used in laboratories has sufficient value to make recovery an economically sound practice. Small amounts of most of the valuable metals can often be recovered and recycled in the laboratory. Such recovery procedures should be carried out by, or under the direct supervision of, a trained professional who understands the chemistry involved. Larger quantities of some metals can often be sold to suppliers or reprocessors for recovery. The original supplier should be contacted for information on names and locations of reprocessors.

A. MERCURY

Metallic mercury should be collected for recovery and recycling. Manipulations with it should be carried out in a hood as much as possible. Small quantities can be made relatively free of insoluble contaminants by filtering a few times through a conical filter paper with a small hole at the bottom of the cone; insoluble contaminants float on top and collect on the sides of the filter cone. The filter paper should be discarded for disposal in a secure landfill. Liquid mercury in quantities greater than 4.5 kg can be sold to a commercial reprocessor; smaller quantities may be accepted at no charge if the donor pays transportation costs. Large quantities can be sent in standard 34.5-kg (76-lb) flasks to reprocessors, who will usually supply the flasks. Most reprocessors will purchase mercury only from institutions, not from individuals.

Much of the metallic mercury from a spill can be collected in a bottle equipped with an eyedropper-type nozzle and connected to a vacuum aspirator. Small droplets can be amalgamated with zinc dust and the resulting solid swept up. Droplets in floor crevices from spills, or in crevices in metal containers, can be converted to mercuric sulfide by dusting with sulfur powder. There are commercial kits that contain materials and small equipment for cleaning up minor spills of mercury. Mercury can be recovered from its salts by converting them to the sulfide. Both amalgams and mercury sulfide are accepted by reprocessors for recovery.

B. SILVER

A procedure developed for recovery of silver and mercury salts from solutions remaining from chemical oxygen demand (COD) tests[1] can be used to recover metallic silver from aqueous solutions of its salts.

$$Ag^+ + NaCl \longrightarrow AgCl + Na^+$$

$$AgCl \xrightarrow{\quad Zn, H_2SO_4 \quad} Ag + \text{soluble Zn salts}$$

The solution of silver salt is acidified with 6 M nitric acid (pH about 2, pH paper) and treated with a 10% excess of 20% aqueous sodium chloride solution. The precipitated silver chloride is collected on a Büchner funnel and washed twice with warm 4 N sulfuric acid and twice with distilled water. The silver chloride is dried and ground to a fine powder. One hundred grams of the dry silver chloride powder is mixed thoroughly with 50 g of pure granulated zinc metal and the mixture stirred with 500 mL of 4 N H_2SO_4. (CAUTION: This operation should be performed in a hood away from any source of ignition because hydrogen is evolved.) When the zinc has dissolved, the supernatant solution is decanted, and the crude silver is again mixed with 50 g of granulated zinc and treated with 500 mL of 4 N H_2SO_4. After the zinc has dissolved, about 5 mL of concentrated H_2SO_4 is added carefully, and the mixture is heated to 90°C with stirring for a few minutes. The silver is separated by filtration and washed with distilled water until the washings are free of sulfate ($BaCl_2$ test). A sample of the resulting silver should give a clear solution in concentrated nitric acid. If the solution is turbid, indicating the presence of silver chloride, the treatment with zinc and 4 N H_2SO_4 should be repeated.

This procedure should give silver of about 99.9% purity. The zinc salt solutions can be flushed down the drain if local regulations permit, or the zinc can be precipitated as the carbonate for landfill disposal.

Scrap silver-based photographic film and photographic developer can be recovered by commercial reprocessors; see Smelters and Refiners—Precious Metals or Gold, Silver, and Platinum Dealers in the Yellow Pages of any large city telephone directory. It is possible to purchase electrolytic units that are designed specifically for recovery of silver from photographic solutions. Silver can also be recovered from photographic solutions by passing them through a cartridge (these are commercially available[2]) filled with steel wool or iron particles. Metal exchange occurs, and the silver drops as a sludge through a screen at the bottom of the cartridge into a liquid-filled space in the cartridge container. The dilute iron-containing effluent can be poured down the drain. This procedure is low cost and reduces the silver content of a solution to 20 mg/L or less. It works best when used on a continuous basis rather than intermittently. For further details on recovery of silver from photographic solutions, see Reference 2.

C. OTHER VALUABLE METALS

The metal values in spent catalysts or solutions that contain such noble metals as platinum, palladium, rhodium, and ruthenium are almost always worth recovering. Generally, the original supplier will accept quantities greater than a specified minimum for credit and recovery. The supplier should be consulted for information on minimum quantities and any pretreatment that may be required.

Small quantities can be recovered in the laboratory by chemical procedures. However, the chemistry of these metals is sufficiently distinct that no general procedure is applicable, and it is best to use procedures for each that are described in the literature. Procedures for recovering platinum from catalysts are given in *Organic Syntheses*[3] and *Inorganic Syntheses*.[4] The chemistry of each of the noble metals, including recovery procedures, can be found in either of two standard references.[5,6]

III. RECOVERY OF SOLVENTS BY DISTILLATION

If the work in a laboratory is of such a nature that used solvents of known composition are routinely produced, solvent recovery may be economically feasible. The costs of recovery should be balanced against the combined costs of purchasing new solvents and of disposing of the used ones. Analysis of the laboratory solvent use pattern can reveal solvents that are used in sufficient quantity to warrant segregation of used material for recovery. Some commercial firms recover solvents in the quantities that are used in laboratories of moderate size.

It can also be feasible to recover many common organic solvents by distillation in laboratory bench-top stills operated by students or technicians under proper supervision. Some laboratory equipment companies offer laboratory-size stills designed for solvent recovery. Larger recovery operations can be conducted in commercial solvent stills; a 20-L/h solvent still can be obtained for about $5000 (1982). Special care should be exercised in distillation recovery of solvents that boil near room temperature to ensure that the condenser and its coolant are capable of preventing significant quantities of uncondensed vapor from escaping into the atmosphere. Peroxide-forming solvents such as ethers should be excluded from any distillation recovery operation. A still used for recovery of ethanol may require a permit from the Federal Bureau of Alcohol, Tobacco, and Firearms, which has offices in larger cities.

The key to successful distillation and recovery is proper segregation of the used solvents at the point of generation. Each solvent should be kept separate, and solvents for recovery should not be highly contaminated. The laboratory waste-disposal plan should identify those solvents that are to be recovered and should indicate the markings on the containers that are to be used for their collection.

A potential problem with recovering solvents is acceptance of the recovered solvent by chemists or other users who may be concerned about its purity. This problem can often be overcome by a pilot trial with the recovered solvents. With proper segregation at the point of generation and careful distillation procedures, solvents of any desired degree of purity can usually be produced. Recovered solvents of relatively low or questionable purity can be used as thinners, solvents, and degreasers by other academic or operational departments of the institution, provided that it is first determined by analysis or from knowledge of the source that the solvent does not contain hazardous contaminants.

IV. EXCHANGE OF UNNEEDED CHEMICALS

Chemicals that are no longer needed by one laboratory may be usable by another laboratory or unit of the same institution. Some laboratories have set up a retained-chemicals storage. Partially used reagents in their original containers are sent to this storage, indexed by card file or computer, and stored on indexed shelves for easy retrieval. Laboratory personnel who need a certain reagent learn to check this storage for the reagent before putting in a purchase order, because the chemicals are supplied free of charge. Some laboratories accept compounds that have been synthesized in their operations into such a system, with the requirement that the sample be properly labeled as to chemical identity, name of the worker who synthesized it, date, and notebook reference.

It is essential that a retained-chemicals storage be in a well-ventilated area equipped with a sprinkler system. It should be inspected regularly. Samples that show evidence of deterioration or on which labels have become illegible should be discarded. Some samples can be saved by relabeling. If the index of retained chemicals is in a computer, and the number of chemicals is not too large, a list of the available chemicals can be distributed periodically to all units of the organization. The retained-chemicals storage can also be a clearinghouse for standing requests for excess chemicals. Some institutions have found that they get standing requests from their laboratories for certain chemicals that become unneeded in their other laboratories.

Arrangements may also be possible for exchange of unneeded chemicals among institutions. For example, exchange, donation, or sale at a nominal price of unopened containers might be arranged between laboratories that are located near each other or among academic institutions that operate under a common administration. The possibility of disposing of unneeded chemicals through a regional chemical exchange should also be considered.

Industrial laboratories sometimes donate unneeded chemicals to nearby academic institutions, and this practice is to be encouraged. However, to avoid loading the recipient with unneeded chemicals, it is best to have a staff member of the academic institution visit to select those reagents that the institution can use. Remember that chemicals being exchanged between institutions are subject to DOT regulations on packing, labeling, and transporting (see Chapter 8).

It can be feasible to exchange or donate unopened or partially used samples of chemicals that are in their original containers, provided that

1. The donor has a control system that gives reasonable assurance of the identity of the chemical, and
2. The recipient accepts full responsibility for the identity and quality of such chemicals.

A retained-chemicals storage or exchange should not accept chemicals that are capable of forming dangerous peroxides (Appendix I, Table I.1, classes 1–7) unless they are known to contain inhibitors.

V. USE OF UNNEEDED CHEMICALS AS FUEL

The heat of combustion of certain types of organic chemicals can be partially recovered by adding them in small quantities to the fuel feed of the institution's power- or steam-generating unit. In general, this procedure can be applied to hydrocarbons and oxygen- or nitrogen-containing compounds. Compounds that contain halogen or sulfur should not be so burned, except as minor contaminants of an acceptable chemical type. Absolutely excluded should be highly toxic compounds, explosives, compounds with a low flash point, and organometallics. Procedures for this method of disposal are given in Chapter 9.

REFERENCES

1. S. Aslam and O. L. Walker, *J. Water Pollut. and Control Fed.* *54*, 1148 (1982).
2. *Silver Recovery Techniques, NMA TR4-1982*, National Micrographics Association, 8719 Colesville Road, Silver Spring, Md. 20910.
3. R. Adams, V. Voorhees, and R. L. Shriner, *Org. Syn. Coll.*, Vol. I, p. 463.
4. G. B. Kauffman and L. A. Teter, *Inorg. Syn.*, Vol. VII, p. 232.
5. H. L. Grube, in *Handbook of Preparative Inorganic Chemistry*, 2nd ed., G. Brauer, ed., Academic Press, New York, 1965, Vol. 2, pp. 1560–1605.
6. I. M. Kolthoff and P. J. Elving, eds., *Treatise on Analytical Chemistry*, Wiley-Interscience, New York, 1963, Part II, Vol. 8.

5 Disposal of Chemicals in the Sanitary Sewer System

I. INTRODUCTION

Only a few years ago it was common practice to dispose of most laboratory wastes down the laboratory drain. Today the indiscriminate drain disposal of laboratory chemicals, without regard to quantity or chemical type, is not acceptable. Laboratory drain systems are connected to sanitary sewer systems, and their effluent will eventually go to a sewage treatment plant. Some chemicals can interfere with the proper functioning of sewage treatment facilities. In the laboratory drain system itself, some chemicals can create hazards of fire, explosion, or local air pollution; others can corrode the drain system.

It is essential to recognize that the characteristics and capabilities of wastewater treatment plants vary from one locality to another and that these factors and local regulations govern what types and concentrations of chemicals they can handle. The drain-disposal procedures and rules in the laboratory waste management plan should be based on a thorough knowledge of local regulations pertaining to materials that are acceptable for disposal in the local sanitary sewer system. It may be helpful to provide the wastewater treatment plant with information on the types and quantities of chemicals that the laboratory site plans to put into the sanitary sewer system and on the toxicity and biodegradability of these chemicals; References 1–6 are useful sources of such information.

Subject to local regulations, modest quantities of many common laboratory chemicals, including some that are listed as hazardous by the U.S. Environmental Protection Agency, can be safely and acceptably disposed of down the laboratory drain with proper precautions.

• Drain disposal should be only into a drain system that flows eventually to a wastewater treatment plant and never into a system, such as a storm sewer system, that flows directly into surface water.

• The quantities of chemicals disposed of in the drain must be limited generally to not more than a few hundred grams or milliliters. The disposal should be performed by flushing with at least 100-fold excess water at the sink, so that the chemicals become highly diluted in the wastewater effluent from the laboratory and even more so by the time the effluent reaches the treatment plant.

• The sewer disposal of laboratory wastes by individual laboratory workers or by students in teaching laboratories should be monitored for adherence to guidelines on types of chemicals, quantities and rates, and flushing procedures. Periodic checks by laboratory supervisors and by waste management personnel are advisable.

It is strongly advised, and may be mandatory in some localities, that the laboratory site provide for automatic sampling of its sanitary sewer effluent and that laboratory personnel collect and analyze daily composite samples of this effluent. Such a procedure is important for controlling drain disposal in the laboratory within limits required by local regulations.

An alternative procedure is to segregate at the point of collection those wastes that are suitable for sewer disposal, either as such or after some treatment, such as neutralization, oxidation, or reduction. These collected wastes can then be fed at a low, controlled rate into the sewer system by a trained, authorized individual. The feed rate should be such that no disruption of the sewage treatment facilities will occur and should be determined by prior consultation with the management of these facilities.

• Certain types of chemicals are permissible for drain disposal; others are not.

II.   DRAIN DISPOSAL OF CHEMICALS

A.   Organic Chemicals

Only those organic compounds that are reasonably soluble in water are suitable for drain disposal. A useful criterion is that developed for qualitative organic analysis: a compound is considered water soluble if it dissolves to the extent of at least 3%, judged by whether 0.2 mL or 0.1 g dissolves in 3 mL of water in a test tube.[7] In general, such materials can be put down the drain, flushed or mixed with at least 100 volumes of excess water. However, this generalization must be tempered with chemical common sense. Substances that boil below 50°C, even though adequately water soluble, should not be poured down the drain because they can cause unacceptable concentrations of vapor in the user's or another laboratory or in the sewer system. An example is diethyl ether, which can create a hazard of fire or explosion. On the other hand, formaldehyde (b.p. −21°C) is an exception because it is so hydrated in water that very little of it vaporizes from dilute aqueous solution. Highly malodorous substances should not be put down the drain.

In general, a substance that meets the solubility criterion, but contains another material that does not, should not be poured down the drain. However, if the water-insoluble material comprises less than about 2% of the mixture, drain disposal is usually acceptable because the small quantity of water-insoluble material will be well dispersed in the aqueous effluent. Common examples are acetone that has been used to rinse glassware and ethanol that contains a hydrocarbon denaturant.

Some organometallic compounds such as Grignard reagents and alkyllithiums or aryllithiums can be decomposed by water into innocuous solutions that are suitable for drain disposal. This decomposition should always be performed in the laboratory before drain disposal (see Chapter 6, Section IV.A) and never in the drain itself. The solubility criterion excludes from drain disposal hydrocarbons, halogenated hydrocarbons, nitro compounds, mercaptans, and most oxygenated compounds that contain more than five carbon atoms. Other exclusions are explosives, such as azides and peroxides, and water-soluble polymers that could form intractable gels in the sewer system. No more than unavoidable traces of highly toxic organic chemicals, whether of synthetic or biological origin, should be allowed into the sewer system.

General guidelines on the types of organic compounds that are suitable for drain disposal are given in Appendix K. In general, compounds that are not listed in Appendix K are *not* suitable.

B. Inorganic Chemicals

Drain disposal is permissible for dilute solutions of inorganic salts in which both cation and anion are listed in the second columns (relatively nontoxic) of Tables 6.1 and 6.2 in Chapter 6. Salts in which either the cation or anion is listed in the first columns of these tables are not permitted in more than unavoidable traces. Mineral acids and alkalies are preferably neutralized before drain disposal. Some laboratories permit drain disposal of these acids and bases by flushing them down with excess water at rates not to exceed 50 mL/minute of the concentrated acid or base, but this practice is falling into disfavor and in any case should be used only in laboratories that have chemically resistant drain systems.

III. BIOLOGICAL PRETREATMENT OF LABORATORY WASTES

Some laboratory wastes that are not acceptable in the sanitary sewer system because of low solubility or chemical type can be biodegraded in a pretreatment operation into material that is acceptable.

One possibility is biological treatment of laboratory wastes in a local industrial wastewater treatment plant. The laboratory waste manager can learn if any local industries have their own pretreatment or treatment facilities and if they are treating waste streams similar to some of the wastes being generated in the laboratory. Because of the relatively small quantity of the laboratory waste, the industry may be willing to accept it into their treatment process. This practice has been used by some universities, which deliver the waste to the industrial treatment plant, where it is slowly added to the industrial waste stream.

An alternative for laboratories that generate modest amounts of some wastes that are not acceptable in the local sanitary sewer system is to set up an in-house, laboratory-scale biological treatment facility.[8] This option is most likely to be useful to laboratories in academic institutions, where knowledge applicable to the design and operation of such a facility should exist and where its operation can be a useful part of the educational program in addition to reducing the volume of hazardous waste. A small-scale biological treatment facility can pretreat many types of organic wastes so that they can become acceptable for sanitary sewer disposal. Examples include phenols, alcohols, aldehydes, ketones, and waste from life-science laboratories.

Many types of biological treatment processes have been developed by municipalities and industries, and these can be duplicated on a small scale in the laboratory. The treatment system can be either aerobic or anaerobic, although most wastewater treatment systems are aerobic, which affects oxidative degradation of organic substrates. Aerobic systems are generally designed as activated sludge or trickling filter systems.[8] Bacteria for an aerobic pretreatment system can be obtained from the local wastewater treatment plant and gradually acclimated to live on and degrade a specific waste. The pretreatment facility can be set up on a scale commensurate with the volume of waste to be treated, for example, a 22-L round-bottomed flask, a fish aquarium, or a livestock watering tank. Such reactors can be operated on either a batch or a continuous basis. The rate at which they can biodegrade wastes depends on the chemical nature of the waste and the degree to which the bacteria are acclimated to this type of chemical.[2,8]

Anaerobic biodegradation is generally slower than aerobic but can be useful for certain wastes that are not readily degraded in aerobic systems. Examples include aromatic hydrocarbons and their chlorinated derivatives. Laboratory-scale anaerobic digesters can be constructed from materials as simple as a 5-gallon (18.9-L) carboy wrapped with heating tape and insulation.[8]

References 9–14 provide information on the theory, design, and operation of biological wastewater treatment systems.

REFERENCES

1. N. I. Sax, *Dangerous Properties of Industrial Materials*, 4th ed., Van Nostrand Reinhold, New York, 1975.
2. W. W. Eckenfelder, Jr., *Principles of Water Quality Management*, CBI, Boston, Mass., 1980.
3. *Toxic and Hazardous Industrial Chemicals Safety Manual*, The International Technical Information Institute, Tokyo, Japan, 1979.
4. *Fate of Priority Pollutants in Publicly Owned Treatment Works*, EPA Office of Water and Waste Management, Washington, D.C., 1981.
5. N. Nemerow, *Industrial Water Pollution*, Addison-Wesley, Reading, Mass., 1978.
6. *Merck Index*, Merck & Co., Rahway, N.J., 1976.
7. R. L. Shriner, R. C. Fuson, D. Y. Curtin, and T. C. Morrill, *The Systematic Identification of Organic Compounds*, 6th ed., John Wiley & Sons, New York, 1980, pp. 90–112.
8. J. Meister, *Operation of the Southern Illinois University (Carbondale) Hazardous Waste Program*, Conference on Waste Management in Universities and Colleges, 1980, Association of Physical Plant Administrators, Suite 250, 11 Dupont Circle, Washington, D.C. 20036; *Proceedings of the 13th Mid-Atlantic Conference on Industrial Waste*, Ann Arbor Press, Ann Arbor, Mich.
9. *Wastewater Treatment Plant Design*, Water Pollution Control Federation, Lancaster Press, Lancaster, Pa., 1977.
10. *Wastewater Treatment Plant Operations, A Manual of Practice*, 4th ed., Water Pollution Control Federation, Lancaster Press, Lancaster, Pa., 1979.
11. *Recommended Standards for Sewage Works*, 1978 ed., Great Lakes Upper Mississippi River Board of State Sanitary Engineers, Health Education Service, Inc., Albany, N.Y., 1978.
12. *Manual of Instruction for Sewage Treatment Plant Operators*, New York Department of Health, Health Education Service, Albany, N.Y., 1979.
13. C. Sawyer and P. L. McCarthy, *Chemistry for Environmental Engineering*, McGraw-Hill, New York, 1978.
14. N. L. Nemerow, *Industrial Water Pollution*, Addison-Wesley, Reading, Mass., 1978.

6 Procedures for Laboratory Destruction of Hazardous Chemicals

CONTENTS OF CHAPTER 6

I. INTRODUCTION

From a chemical point of view it is feasible to reduce or destroy the hazard characteristic of many hazardous chemicals by chemical reaction in the laboratory. Although in-house chemical destruction of chemicals is not likely to be an economically practical general solution to the disposal of laboratory wastes, many laboratories may find the route useful for certain wastes. Academic laboratories may find that laboratory destruction of some chemicals can be made an effective part of their instructional programs, while at the same time reducing the quantities of wastes that have to be disposed of off site.

This chapter is an initial effort in that direction. It presents guidelines and procedures for destroying and disposing of common classes of laboratory chemicals. Literature sources for some of the procedures presented are given in References 1–6. Information on specific compounds that may be included in this chapter only as members of a class should be sought in References 1–4.

Several destruction procedures are presented in the detail usually reserved for synthetic procedures. It would be helpful to the chemical community if chemists were to work out details of additional destruction procedures, write them up in the style of *Organic Syntheses*, have them checked experimentally in another laboratory for efficacy and safety, and have them published. The International Agency for Research on Cancer (IARC) has begun a series of monographs on procedures for destroying carcinogens and has published the first two, on aflatoxins[7] and on *N*-nitrosamines.[8] Similar monographs are planned on *N*-nitrosamides, polycyclic aromatic hydrocarbons, alkylating agents, halogenated compounds, aromatic amines, and hydrazines.

It is a fundamental principle of good laboratory practice that no one should undertake manipulations with any chemical without first understanding the properties and possible hazards of that chemical. The destruction/disposal procedures presented here are intended to be carried out only by, or under the direct supervision of, a trained professional who understands the chemistry involved. The laboratory practices recommended in Section I of *Prudent Practices for Handling Hazardous Chemicals in Laboratories*,[3] should be followed, making proper use of laboratory fume hoods.[9] The procedures are intended for application to laboratory quantities—not more than a few hundred grams at a time. Because hazards tend to increase exponentially with scale, larger quantities should be treated only in batches of this size unless a qualified chemist in the laboratory concerned has demonstrated that the procedure can be scaled up safely. Some laboratories have prepared handbooks on the properties, health hazards, and disposal methods of compounds that they use (for an example, see Reference 10).

II. ORGANIC CHEMICALS

Most organic compounds can be destroyed in properly designed and operated incinerators (see Chapter 9), and the chemical classes that are suitable for this method of destruction are indicated in the following sections. However, many laboratories do not have their own incinerator and find that commercial incinerator operators either are not willing to accept laboratory waste or are too expensive; accordingly, they dispose of their hazardous waste in a secure landfill (see Chapter 10).

Some chemicals can be disposed of in sanitary sewage systems (see Chapter 5) or by chemical destruction in the laboratory. These options will be considered in the following discussion of principal classes of organic chemicals.

A. Hydrocarbons

This class includes alkanes, alkenes, alkynes, and arenes. They burn well and can be disposed of by incineration or as fuel supplements (see Chapter 9). Many of those commonly used in laboratories are classified by the U.S. Environmental Protection Agency (EPA) as hazardous because of the characteristic of ignitability; these can be put into a secure landfill only in the small quantities that can be packaged in a lab pack (see Chapter 10). Because they are virtually insoluble in water, only traces should be put down the drain.

Some alkenes, especially cyclic ones like cyclohexene, may form explosive peroxides on long storage with access to air. Old samples should be tested for peroxides and treated appropriately if they are present (see Section II.P of this chapter).

Some polycyclic arenes, such as benz[*a*]pyrene and 1,2-benzanthracene, are potent animal carcinogens and suspected human carcinogens. Although such hydrocarbons can be incinerated or put into a lab pack for disposal in a secure landfill, they or mixtures containing them must be packaged separately and carefully labeled so that operators can handle them without being exposed. Up to a few grams of polycyclic arenes can be destroyed by being dissolved in a fresh solution of sodium dichromate in concentrated sulfuric acid.[11]

Procedure for Destroying One Gram of a Polycyclic Arene. A 0.27 M solution of sodium dichromate is prepared by dissolving 20 g of

$Na_2Cr_2O_7 \cdot 2H_2O$ in 10 mL of water and gradually adding enough 96% sulfuric acid to bring the volume to 250 mL. One gram of polycyclic arene is gradually added at room temperature with stirring until it is dissolved. The solution is allowed to stand for 2 days. The solution, now free of carcinogen, is carefully poured into 1 L of water with stirring, treated with up to 84 g of sodium bisulfite to reduce excess Cr(VI) to Cr(III), and brought to above pH 7 with sodium hydroxide solution to precipitate the chromium as $Cr(OH)_3$ (Appendix J), which is disposed of as described in this chapter, Section III.B.

This procedure should be used only on a small scale because of the potential hazards from splashing or spilling large quantities of dichromate/sulfuric acid solutions. Small pieces of glassware contaminated with such hydrocarbons can be decontaminated by immersion in such a solution for 2 days.

B. HALOGENATED HYDROCARBONS

A number of halogenated hydrocarbons are common laboratory solvents, for example, methylene chloride, chloroform, carbon tetrachloride, tetrachloroethylene, 1,1,1-trichloroethane, 1,1,2-trichloro-1,2,2-trifluoroethane, chlorobenzene, and o-dichlorobenzene. Other halogenated hydrocarbons are important laboratory reagents, examples being *tert*-butyl chloride, benzyl chloride, ethyl bromide, methyl iodide, bromobenzene, methylene iodide, 1,2-dibromoethane, and allyl chloride. Many of these halogenated hydrocarbons are fairly toxic, indicating the need for prudence in handling and disposing of them in laboratory quantities.

Those members of this class that are used in quantity as solvents are candidates for recovery by distillation. Their low solubility in water makes quantities greater than traces unacceptable in the sanitary sewer. Most laboratories dispose of unrecoverable halogenated hydrocarbons in landfill lab packs. When blended with excess nonhalogenated solvents, they can be destroyed by incineration, but more than minor quantities require an incinerator equipped with a scrubber to remove the hydrogen halide from combustion gases.

A few highly halogenated arenes, such as polychlorinated biphenyls, are not completely destroyed by most incinerators. Only two or three incinerators in the United States have the demonstrated capability and the permits for destruction of these refractory substances. A few secure landfills can accept them.

Up to a few hundred grams of alkyl halides can be destroyed by hydrolysis with ethanolic potassium hydroxide. The general procedure given below is applicable to most primary, secondary, and tertiary alkyl chlorides, bromides, and iodides. It can also be used for allyl and benzyl halides, including fluorides, and for compounds that have two or even three chlorine, bromine, or iodine atoms on one carbon. It is not useful for vinyl or aryl halides or for most alkyl fluorides.

The procedure can also be used to hydrolyze the following classes of alkylating agents, all of which contain some members that are human or animal carcinogens: dialkyl sulfates, alkyl alkanesulfonates and arenesulfonates, chloromethyl ethers, epoxides, and aziridines. Although some of the more active alkylating agents can be hydrolyzed effectively by treatment with aqueous sodium hydroxide (this chapter, Section II.J), the following procedure is more broadly applicable. It uses 20% excess potassium hydroxide based on the equation

$$RX + KOH \longrightarrow ROH + KX$$

An additional equivalent of potassium hydroxide must be used for each additional hydrolyzable group in the compound being treated. Competing reactions, such as formation of ethyl ethers (ROC_2H_5) or dehydrohalogenation, also destroy the organic halide. Dehydrohalogenation becomes the principal reaction with tertiary halides, generating olefins.

$$t\text{-}C_4H_9Cl + KOH \longrightarrow (CH_3)_2C=CH_2 + KCl + H_2O$$

(CAUTION: Reactions that may generate olefins too volatile to be trapped by a reflux condenser, such as isobutylene or 2-methyl-2-pentene, should be carried out in a good hood away from any source of ignition.) Although a few hydrolysis products, such as an aldehyde formed from a 1,1-dihalide, may undergo condensation reactions, this does not interfere with the destruction of the organic halide.

Procedure for Hydrolyzing 1.0 Mole of an Alkyl Halide. Place 79 g (1.20 mol) of 85% potassium hydroxide pellets in a 1-L three-necked flask equipped with a stirrer, water-cooled condenser, dropping funnel, and heating mantle or steam bath. With brisk stirring, 315 mL of 95% ethanol is added rapidly. The potassium hydroxide dissolves within a few minutes, causing the temperature of the solution to rise to about 55°C.[12] The solution is heated to gentle reflux, and the liquid alkyl halide (or its solution in 95% ethanol if it is a solid) is added dropwise. The rate of addition and the heat input are adjusted to maintain gentle reflux. The dropping funnel is rinsed with a little ethanol, and stirring and reflux are continued for 2 hours (stirring is essential to prevent bumping caused by the potassium halide that often precipitates). If the reaction products are water soluble, the mixture is diluted with 300 mL of water, cooled to room temperature, neutralized, and washed down the drain with about 50 volumes of water. Otherwise, the mixture is neutralized and sent to an incinerator or landfill.

The procedure is most useful for halides that are hydrolyzed to water-soluble products; those that are not can generally be incinerated or sent to a secure landfill. However, for some in the latter category, the procedure may be useful for decomposing a toxic compound or for destroying the hazard characteristic of reactivity (see Appendix A, 40 CFR 261.23), which would otherwise prevent the compound from being put into a landfill. Many organic halides, even though not classed as reactive by the EPA characteristic, can hydrolyze slowly. If they contain moisture when packed for landfill disposal, the hydrogen halide generated by hydrolysis will slowly destroy their bottle caps unless these are made of polypropylene or lined with Teflon.

C. OTHER HALOGENATED COMPOUNDS

This broad class comprises halogenated hydrocarbons that contain a functional group such as hydroxyl, carboxyl, thiol, amino, or nitro. Acid halides are discussed separately (this chapter, Section II.J). Members of this class can be incinerated or, if not reactive (see Appendix A, 40 CFR 261.23), put into a secure landfill. Hydrolysis with ethanolic potassium hydroxide can destroy many of these compounds; those that contain base-reactive groups will require additional potassium hydroxide.

In general, halogen substituents decrease the water solubility of organic compounds, but a few are sufficiently soluble to be flushed down the drain, e.g., 2-chloroethanol and chloroacetic acid. However, some that have good solubility are too toxic for such disposal, e.g., fluoroacetic acid.

D. ALCOHOLS AND PHENOLS

Alcohols are used extensively as solvents and reagents in the laboratory. They can be incinerated, burned as fuel supplement, or put into a secure landfill. Many of the common ones are readily water soluble, have low toxicity, are readily biodegradable, and can be flushed down the drain.

Phenols tend to be more toxic than alcohols. They are skin irritants and some are allergens, so care must be taken to avoid skin contact while disposing of them. They can be incinerated or put into a secure landfill. Many phenols, particularly phenol and its monosubstituted derivatives, are efficiently degraded by biological treatment systems and can be destroyed in an on-site biological treatment facility (see Chapter 5, Section III). Because phenols in water that enters a water-treatment plant are chlorinated to chlorophenols, which give drinking water a bad taste, phenols should never be allowed into storm sewers or other drains that lead directly to streams that are sources of drinking water. Moreover, it is best to avoid disposing of phenols even in drains leading to off-site wastewater-treatment plants because the plant's efficiency may occasionally be low enough to let some phenolic material through to the chlorination system.

Aqueous solutions that contain low concentrations of phenols can be passed through a bed of activated carbon, which adsorbs the phenols. There are small industrial adsorption units that consist of a tank or column filled with carbon granules and that can accommodate flow rates of about 20 L/min. If such units are to be used regularly, facilities for regenerating the carbon can be added. On a smaller scale, a glass column filled with pelletized activated carbon (for example, Norit) can be used to adsorb phenols from solutions. Suppliers of activated carbon can furnish information on the adsorptive capacity of various types of carbons for various compounds. Spent activated carbon can be incinerated or disposed of in a landfill, depending on what it has adsorbed. The effluent solution should contain so little phenolic material that it can be disposed of in the sanitary sewer system.

Small amounts of phenols can be destroyed in the laboratory by hydrogen peroxide in the presence of an iron catalyst[13]; the phenolic ring is ruptured to form harmless fragments. The procedure is illustrated for phenol.

Procedure for Decomposing 0.5 Mole of Phenol. A solution of 47 g (0.50 mol) of phenol in 750 mL of water is prepared in a 2-L three-necked flask equipped with a stirrer, dropping funnel, and thermometer. Ferrous sulfate heptahydrate (23.5 g, 0.085 mol) is then dissolved in the mixture, and the pH is adjusted to 5–6 (pH paper) with dilute sulfuric acid. Then 410 mL (4.0 mol) of 30% hydrogen peroxide is added dropwise with stirring over the course of about an hour. (CAUTION: The order of addition of the reagents is important. If hydrogen peroxide and ferrous sulfate are premixed, a violent reaction may occur.) Heat is evolved, and the reaction temperature is maintained at 50–60°C by adjusting the rate of addition and by using an ice bath if necessary. Stirring is continued for 2 hours while the temperature gradually falls to ambient. The solution is allowed to stand overnight and is then washed down the drain with a large volume of water.

This procedure can be applied to many phenols, including the cresols, monochlorophenols, and naphthols. More water is generally needed to dissolve them; thus, 0.5 mol of a cresol requires about 2.5 L of water.

E. ETHERS

Open-chain monoethers, both aliphatic and aromatic, are relatively non-toxic, and those that are classified hazardous on the EPA lists (see Appendix A, 40 CFR Subpart D) meet only the hazard characteristic of ignitability. Accordingly, they are not candidates for laboratory destruction. Except for diethyl ether, they are not sufficiently water soluble for drain disposal. Even diethyl ether should not be put down the drain because of its low boiling point (35°C). Small amounts of diethyl ether can be evaporated in a hood (with no open flame, hot surface, or other source of ignition in the hood) if the ether is free of peroxides or contains an inhibitor (which most commercial laboratory diethyl ether now does). Diethyl ether can also be mixed with at least 10 volumes of higher-boiling solvents for incineration.

The common cyclic ethers tetrahydrofuran and dioxane are sufficiently water soluble for drain disposal, although this may be prohibited in some localities because of their low flash points. They can be incinerated or put into a secure landfill.

Lower alkyl ethers and ether acetates of diethylene and triethylene glycol (glyme, diglyme, carbitols, cellosolves) are water soluble and can be put down the drain.

Many ethers can form explosive peroxides on exposure to air (see Appendix I). The disposal of peroxide-containing ethers is discussed in this chapter, Section II.P.

Some of the macrocyclic polyethers, "crown ethers," are quite toxic. Small quantities can be destroyed by sodium dichromate in sulfuric acid (this chapter, Section II.A).

Epoxides are potent alkylating agents, and many are highly toxic. The epoxide group can be destroyed by alkaline hydrolysis as described for alkyl halides (this chapter, Section II.B).

F. MERCAPTANS

Mercaptans can be incinerated or put into a secure landfill. The relatively small quantities usually employed in laboratory work can be destroyed by oxidizing the mercaptan group to a sulfonic acid group with sodium hypochlorite. If the mercaptan

$$RSH + 3NaOCl \longrightarrow RSO_3H + 3NaCl$$

contains other groups that can be oxidized by hypochlorite, the quantity of this reagent used must be increased accordingly.

Procedure for Oxidizing 0.5 Mole of a Liquid Mercaptan. Pour 2.5 L (1.88 mol, 25% excess) of commercial laundry bleach (e.g., Clorox) containing 5.25% sodium hypochlorite into a 5-L three-necked flask located in a fume hood. The flask is equipped with a stirrer, thermometer, and dropping funnel. The mercaptan (0.50 mol) is added dropwise to the stirred hypochlorite solution, which is initially at room temperature. Oxidation usually starts soon, accompanied by a rise in temperature and dissolution of the mercaptan. If the reaction has not started spontaneously after about 10% of the mercaptan has been added, addition is stopped and the mixture is heated to about 50°C to get it started. Addition is resumed only after it is clear that oxidation is occurring. The temperature is maintained at 45–50°C by adjusting the rate of addition and using an ice bath if necessary; addition requires about an hour. Stirring is continued for 2 hours while the temperature gradually falls to ambient. The mixture should be a clear solution, perhaps containing traces of oily by-products. The reaction mixture can be flushed down the drain with excess water. The unreacted laundry bleach need not be decomposed. A solid mercaptan can be added gradually through a neck of the flask or can be dissolved in tetrahydrofuran and the solution added to the hypochlorite. Traces of mercaptan can be rinsed from the reagent bottle and the dropping funnel with tetrahydrofuran and the rinsings added to the oxidizing solution.

Calcium hypochlorite is an alternative to sodium hypochlorite and requires only one third as much water. For 0.5 mol of mercaptan, 210 g (25% excess) of 65% calcium hypochlorite (technical grade) is stirred into 1 L of water at room temperature. Most soon dissolves. The mercaptan is then added, as in the above procedure.

Laboratory glassware, hands, and clothing contaminated with mercaptans can be deodorized by a solution of Diaperene, a tetraalkylammonium salt used to deodorize diaper pails.

G. OTHER ORGANOSULFUR COMPOUNDS

Sulfonic acids are discussed below in Section II.I, and sulfonyl halides are discussed in Section II.J. Sulfides, sulfoxides, sulfones, thioamides, and sulfur heterocycles can be incinerated or landfilled. Small amounts of sulfides, RSR, can be oxidized to sulfones, RSO_2R, to eliminate their disagreeable odor. The hypochlorite procedure used for mercaptans (above in Section II.F) can be employed for this purpose, although the resulting sulfones are often water insoluble and may have to be separated from the reaction mixture by filtration.

Carbon disulfide poses a special problem. It is highly volatile (b.p. 46°C), highly toxic, and readily forms ignitable or explosive mixtures with air—even a hot steam line can ignite such a mixture. Incineration is feasible but must be done with care; containers of more than about 100 g of carbon disulfide can cause explosions in incinerators. It should be diluted with about 10 volumes of a higher-boiling solvent before incineration. Its low boiling point, toxicity, and odor preclude landfilling. Small quantities can be evaporated in a hood (with no open flame, hot surface, or other source of ignition in the hood) or destroyed by hypochlorite oxidation as described for mercaptans (see Section II.F above).

$$CS_2 + 8OCl^- + 2H_2O \longrightarrow CO_2 + 2H_2SO_4 + 8Cl^-$$

For each 0.5 mol (38 g, 30 mL) of carbon disulfide, a 25% excess of hypochlorite is used in the form of 6.7 L of laundry bleach or a mixture of 550 g of 65% calcium hypochlorite in 2.2 L of water. The reaction temperature should be regulated at 20–30°C to avoid volatilizing carbon disulfide.

H. CARBOXYLIC ACIDS

Incineration or burial in a secure landfill requires only that the acids be collected in plastic or glass containers because of their corrosiveness to metals. Care must be taken to keep them separate from amines and other bases, with which they form salts with evolution of heat. Water-soluble carboxylic acids and their sodium, potassium, calcium, and magnesium salts can be washed down the drain if local regulations permit.

I. OTHER ORGANIC ACIDS

Sulfonic acids, RSO_3H, are generally soluble in water and can usually be handled according to the guidelines for carboxylic acids: incinerate, use the drain if local regulations permit, or put them in a secure landfill.

Acids of nonhazardous elements, such as phosphonic acids $[RPO(OH)_2]$ and boronic acids $[RB(OH)_2]$, can be handled like sulfonic acids. Acids of hazardous elements, such as arsonic acids $[RAsO(OH)_2]$ and stibonic acids $[RSbO(OH)_2]$, should be sent to secure landfills.

J. ACID HALIDES AND ANHYDRIDES

Acyl halides such as CH_3COCl and C_6H_5COCl, sulfonyl halides such as $C_4H_9SO_2Cl$ and $C_6H_5SO_2Cl$, and anhydrides such as $(CH_3CO)_2O$ react readily with water, alcohols, and amines. They should never be allowed to come into contact with wastes that contain such substances. Moreover, most of them are sufficiently water reactive that they cannot be put into lab packs for landfill disposal. Incineration presents no problems other than those associated with halogen-containing compounds.

Most compounds in this class can be hydrolyzed to water-soluble products of low toxicity that can be flushed down the drain.

Procedure for Hydrolyzing 0.5 Mole of $RCOX$, RSO_2X, or $(RCO)_2O$.

$$RCOX + 2NaOH \longrightarrow RCO_2Na + NaX + H_2O$$

A 1-L three-necked flask equipped with a stirrer, dropping funnel, and thermometer is placed on a steam bath in a hood. Then 600 mL of a 2.5 N sodium hydroxide solution (1.5 mol, 50% excess) is poured into the flask. A few milliliters of the acid derivative is added dropwise with stirring. If reaction occurs, as indicated by a rise in temperature and solution of the acid derivative, addition is continued without heating the mixture. If reaction is sluggish, as may be the case with larger molecules, such as *p*-toluenesulfonyl chloride, the mixture is heated to about 90°C before adding any more acid derivatives. When the initially added material has dissolved, the remainder is added dropwise. When a clear solution has been obtained, the mixture is cooled to room temperature, neutralized to about pH 7 with dilute hydrochloric or sulfuric acid, and washed down the drain with excess water.

If the acid derivative is a solid, it can be added in small portions through a neck of the flask.

K. OTHER ACID DERIVATIVES: ESTERS, AMIDES, NITRILES

These compounds can be incinerated or put into a secure landfill.

Only a few carboxylic esters are sufficiently toxic to require special handling. β-Propiolactone is an acylating agent that has shown potent carcinogenicity in skin-painting tests on animals. However, it is readily hydrolyzed to the relatively harmless β-hydroxypropionic acid with aqueous sodium hydroxide (see Section II.J above), and this hydroxy acid can be safely flushed down the drain. Acrylic esters, which are allergenic to some people, can be hydrolyzed to relatively harmless acrylic acid by ethanolic potassium hydroxide (this chapter, Section II.B); methacrylates, which are less toxic, can be treated the same way.

In contrast to carboxylic esters, sulfonic acid esters and alkyl sulfates are powerful alkylating agents, and some have shown carcinogenicity to animals. They can be readily decomposed by ethanolic potassium hydroxide (this chapter, Section II.B). Alkyl phosphates and phosphonates can be hydrolyzed by the same procedure.

Carboxamides, including the common solvents N,N-dimethylformamide and N,N-dimethylacetamide, are not very hazardous if good laboratory practices are followed. They can be hydrolyzed by refluxing for 5 hours in 250 mL of 36% hydrochloric acid per mole of amide. Hexamethylphosphoramide (HMPA), a potent carcinogen in animal tests, can be treated the same way; one uses 750 mL of 36% hydrochloric acid per mole of HMPA. The analytical reagent thioacetamide, a moderately active carcinogen in animal tests, is decomposed by 25% excess hypochlorite as described for mercaptans (this chapter, Section II.F).

$$CH_3CSNH_2 + 4OCl^- + H_2O \longrightarrow CH_3CONH_2 + H_2SO_4 + 4Cl^-$$

It is sometimes suggested that nitriles be converted into water-soluble and less toxic substances by treating them with sodium hypochlorite. Although this treatment is effective for inorganic cyanides, it is not effective for organic nitriles. Nitriles can be converted into carboxylic acids by refluxing with excess ethanolic potassium hydroxide for several hours (this chapter, Section II.B) or with 250 mL of 36% hydrochloric acid per mole of nitrile for 5–10 hours. Acrylonitrile, a highly toxic compound that should never be put down the drain, can be hydrolyzed by either method.

L. ALDEHYDES AND KETONES

Ketones, especially acetone and methyl ethyl ketone, are very common laboratory solvents. Many aldehydes and ketones are frequently used intermediates in organic synthesis. All of them burn easily and can be incinerated or burned as fuel supplements.

Many aldehydes are irritating to breathe, and some, such as formaldehyde and acrolein, are quite toxic. There is sometimes merit in oxidizing a waste aldehyde to the corresponding carboxylic acid, which is usually a less toxic, less volatile substance. This can be done with aqueous potassium permanganate.

$$3RCHO + 2KMnO_4 \longrightarrow 2RCO_2K + RCO_2H + 2MnO_2 + H_2O$$

Procedure for Permanganate Oxidation of 0.1 Mole of Aldehyde.[14] A mixture of 100 mL of water and 0.1 mol of aldehyde is stirred in a 1-L round-bottomed flask equipped with a thermometer; dropping funnel; stirrer; steam bath; and, if the aldehyde boils below 100°C, a condenser. About 30 mL of a solution of 12.6 g (0.08 mol, 20% excess) of potassium permanganate in 250 mL of water is added over a period of 10 minutes. If this addition is not accompanied by a rise in temperature and loss of the purple permanganate color, the mixture is heated by the steam bath until a temperature is reached at which the color is discharged. The rest of the permanganate solution is added at this temperature ± 10°C. The temperature is then raised to 70–80°C, and stirring is continued for an hour or until the purple color has disappeared, whichever is first. The mixture is cooled to room temperature and acidified with 6 N sulfuric acid. (CAUTION: Do not add concentrated sulfuric acid to permanganate solution because explosive Mn_2O_7 may precipitate.) Enough solid sodium bisulfite (at least 8.3 g, 0.08 mol) is added with stirring at 20–40°C to reduce all the manganese to the divalent state, as indicated by loss of purple color and dissolution of the solid manganese dioxide. The mixture is washed down the drain with a large volume of water.

If the aldehyde contains a double bond, as in the case of the highly toxic acrolein, 4 moles (20% excess) of permanganate per mole of aldehyde is used to oxidize the double bond as well as the aldehyde group.

Most ketones have such low toxicity and are so easily handled that there is little incentive to convert them into something else for disposal. In special cases it may be desirable to decompose one. For example, methyl ketones, such as the neurotoxin methyl *n*-butyl ketone, can be converted into carboxylic acids with laundry bleach (5.25% NaOCl), using an *Organic Syntheses* procedure as a model.[15]

$$RCOCH_3 + 3NaOCl \longrightarrow RCO_2Na + CHCl_3 + 2NaOH$$

An unsaturated methyl ketone, such as methyl vinyl ketone, can be destroyed the same way. In general, most unsaturated ketones can be broken apart at the double bond by the procedure for oxidizing aldehydes given above but using 3.2 moles (20% excess) of permanganate per mole of ketone.

$$3RCOCH{=}CHR + 8KMnO_4 \longrightarrow$$
$$3RCOCO_2K + 3RCO_2K + 2KOH + 8MnO_2 + 2H_2O$$

M. AMINES

Aliphatic and aromatic amines are used principally as synthesis intermediates. A few tertiary amines, such as pyridine and triethylamine, are used as solvents or catalysts. They can be incinerated or put into a secure landfill; they should not be mixed with waste acids.

Aromatic amines are relatively toxic compounds; many of them have an adverse effect on hemoglobin. Some, especially among those with more than one aromatic ring, have shown carcinogenicity to animals and in some cases to humans (e.g., 2-naphthylamine, 4-aminobiphenyl). Aromatic amines can be deaminated to the corresponding arenes, which are less toxic, by diazotization followed by reduction with hypophosphorous acid.[16]

$$ArNH_2 + HNO_2 + HCl \longrightarrow ArN_2Cl + 2H_2O$$
$$ArN_2Cl + H_3PO_2 + H_2O \longrightarrow ArH + H_3PO_3 + HCl + N_2$$

The following procedure uses a greater excess of nitrous and hypophosphorous acids than most reported diazo reactions because the aim is complete deamination rather than the synthesis of a pure product in high yield.

Procedure for Deamination of 0.2 Mole of a Primary Aromatic Amine. To a 1-L three-necked flask equipped with stirrer, thermometer, and dropping funnel are added 25 mL of water, 75 mL of 36% hydrochloric acid, and 0.2 mol of a primary aromatic amine. The temperature is maintained at -5 to $0°C$ by a cooling bath, while 15 g (0.211 mol) of 97% sodium nitrite dissolved in 35 mL of water is added dropwise to the solution or slurry of amine hydrochloride. Stirring is continued an additional 30 minutes after addition is complete. While maintaining the temperature at -5 to $0°C$, 416 mL (4.0 mol) of 50% hypophosphorous acid (precooled to $0°C$) is added over 10–15 minutes. Stirring is continued for 1 hour. The mixture is allowed to stand at room temperature for 24 hours and is then extracted with two 100-mL portions of toluene. The toluene extract of the deamination product is sent to an incinerator or landfill, and the aqueous phase is washed down the drain with excess water.

N. Nitro Compounds

Both aliphatic and aromatic nitro compounds are relatively toxic, and both types can form explosive compounds on exposure to strong bases. Some polynitroaryl compounds are powerful explosives in their own right, trinitrotoluene and picric acid (2,4,6-trinitrophenol) being well-known examples.[17] Picric acid, a common laboratory reagent, is discussed in Chapter 7, Section II.

Most nitro compounds, except those that are known or suspected to be explosive, should be incinerated or landfilled. Explosives should be disposed of only by people trained in the handling of such compounds, who generally detonate them under carefully controlled conditions (see Chapter 7).

O. *N*-Nitroso Compounds

This class is the subject of considerable research because of the carcinogenicity of some of its members, so there is great interest in good methods of destroying small amounts. Incineration is an effective means of destruction. The International Agency for Research on Cancer (IARC) has published a monograph with detailed directions for laboratory destruction of *N*-nitrosamines in laboratory waste.[8] For destruction of milligram quantities or for removing traces from the surface of glassware, a 3% solution of hydrobromic acid in acetic acid at room temperature is effective. For destroying gram quantities, reduction to the amine by aluminum-nickel alloy in alkali works well.

Denitrosation of One Milligram of a Dialkylnitrosamine with HBr.

$$R_2NNO + HBr \longrightarrow R_2NH + BrNO$$

In biological research one may have to dispose of an extract containing a few milligrams of a dialkylnitrosamine in an inert solvent, particularly methylene chloride. A methylene chloride solution containing about 1 mg of a dialkylnitrosamine is concentrated to 1–2 mL, dried over anhydrous sodium sulfate (water makes the HBr reagent less effective), and mixed with 5 mL of a 3% solution of HBr in acetic acid. The latter solution is prepared by diluting commercial 30% HBr in acetic acid with glacial acetic acid. The mixture is kept at ambient temperature for 2 hours. It is then diluted with water, neutralized with 10% aqueous sodium hydroxide, and flushed down the drain with a large excess of water.

Decomposition of 0.05 Mole of a Dialkylnitrosamine with Aluminum-Nickel Alloy.

$$R_2NNO \xrightarrow[\text{NaOH}]{\text{Al}-\text{Ni}} R_2NH$$

This procedure is applicable to dialkylnitrosamines that are more than 1% soluble in water, which generally means those that contain less than 7 carbon atoms. A solution of 0.05 mol of the dialkylnitrosamine in 500 mL of water is prepared in a 2-L three-necked flask that is located in a hood and equipped with a stirrer, an ice bath, and an outlet tube leading through plastic tubing to the back of the hood to carry off the hydrogen that is evolved. Then 500 mL of 1 M sodium hydroxide is added, followed by 50 g of 50/50 aluminum-nickel alloy powder gradually added in small portions over a period of about an hour; the flask neck through which it is added is stoppered between additions. The reaction is highly exothermic and accompanied by frothing that can cause the reaction mixture to foam out of the flask if the alloy is added too rapidly. Stirring is continued for 3 hours in the ice bath and then for 20 hours at ambient temperature. The finely divided black nickel is allowed to settle, and the aqueous phase is then decanted, neutralized, and flushed down the drain with excess water.

The residual nickel can ignite in air if allowed to dry. Therefore, it is washed with 100 mL of water, suspended in 200 mL of water in the three-necked flask, and 800 mL of 1 N hydrochloric acid is added gradually with stirring, which is continued until the nickel has dissolved. The resulting solution of nickel chloride is treated as described in Section III.B of this chapter.

This procedure can be used to destroy dialkylnitrosamines that are less than 1% soluble in water by dissolving the 0.05 mol of the dialkylnitrosamine in 500 mL of methanol instead of water.

Small quantities of *N*-nitrosamides, such as *N*-nitrosomethylurea and *N*-methyl-*N'*-nitro-*N*-nitrosoguanidine (MNNG), can be destroyed by the hydrobromic acid procedure described for *N*-nitrosodialkylamines. However, the aluminum-nickel alloy procedure cannot be used with *N*-nitrosamides because on exposure to base they evolve diazomethane, which is highly toxic and explosive.[18]

P. ORGANIC PEROXIDES AND HYDROPEROXIDES

This group includes dialkyl peroxides (e.g., t-$C_4H_9OOC_4H_9$-t), diacyl peroxides [e.g., $C_6H_5C(O)OOC(O)C_6H_5$], and hydroperoxides (e.g., t-C_4H_9OOH), which are used primarily as initiators for free-radical reactions. However, many substances can form peroxides on storage in contact with air, and safe disposal of such unwanted peroxides is a more common problem than the disposal of excess peroxide reagents.

Jackson et al.[19] and Burfield[20] have published excellent articles on the handling and control of peroxidizable compounds. Several common classes of organic compounds can form peroxides on storage (see Appendix I). The peroxides are sensitive to heat, friction, impact, and light and are among the most hazardous chemicals that are encountered in laboratories. Their hazard potential is all the greater because they may not be suspected or detected in commonly used solvents and reagents. Many explosions have occurred during distillation of peroxide-containing substances, particularly when the distillation has been taken to or near to dryness. Laboratory operations with peroxides or peroxide-containing solvents should always be carried out behind shields and with other precautions.[21]

Peroxidizable compounds should be tested for peroxide content before they are used. The simplest qualitative test is based on oxidation of iodide ion to iodine by the peroxide. One adds 1 mL of the substance to be tested to a freshly prepared solution of 100 mg of sodium or potassium iodide in 1 mL of glacial acetic acid. A yellow color indicates a low concentration of peroxide in the sample; a brown color indicates a high concentration. The test is sensitive for hydroperoxides (ROOH), which constitute the principal hazards of peroxidizable solvents, but does not detect difficultly reducible peroxides such as dialkyl peroxides (ROOR). The latter can be detected by a reagent prepared by dissolving 3 g of sodium iodide in 50 mL of glacial acetic acid and adding 2 mL of 37% hydrochloric acid.[22] Test paper containing a peroxidase is available that detects organic peroxides (including dialkyl peroxides) and oxidizing anions (e.g., persulfate, chromate) colorimetrically.[23]

All peroxidizable solvents and reagents should be dated at the time they are first opened for use. These materials should be discarded or tested for peroxides within a fixed time after opening (see Reference 19 and Appendix I).

Peroxides can be removed from a solvent by passing it through a column of *basic* activated alumina,[24] by treating it with indicating molecular sieves,[20] or by reduction with ferrous sulfate. Although these procedures remove hydroperoxides, which are the principal hazardous contaminants of peroxide-forming solvents, they do not remove dialkyl peroxides, which are usually present in small quantities. A 2-cm × 33-cm column filled with 80 g of 80-mesh F-20 Alcoa or Woelm basic activated alumina is usually sufficient to remove all peroxides from 100–400 mL of common peroxidizable liquids, whether water soluble or water insoluble. After passage through the column the solvent should be re-tested to confirm that peroxides have been removed. Peroxides formed by air oxidation are usually decomposed by the alumina, not merely adsorbed on it. However, to be on the safe side the wet alumina should be slurried with a dilute acidic solution of ferrous sulfate before being discarded or should be put in a capped plastic jar or a polyethylene bag for incineration.

A promising new method for removing peroxides from ethers involves refluxing 100 mL of the ether with 5 g of 4–8 mesh, indicating activated 4A molecular sieve pellets for several hours under nitrogen.[20] The sieve pellets that are separated from the ether present no potential hazard because hydroperoxides are destroyed during the operation, probably by catalytic action of the indicator in the pellets.

Peroxides in 1 L of a water-insoluble solvent can usually be removed by treatment with a solution of 6 g of $FeSO_4 \cdot 7H_2O$ and 6 mL of concentrated sulfuric acid in 11 mL of water. This solution and the solvent are shaken in a separatory funnel or stirred vigorously in a flask until the solvent no longer gives a positive test for peroxide; often a few minutes will suffice.

Peroxide-containing solvents can often be safely incinerated, particularly if the peroxide content is low. Such solvents should not be mixed with other waste solvents but kept in their original containers, which are incinerated one at a time. If the solvent contains a high concentration of peroxide, it should first be diluted with the same unperoxidized solvent or with a common solvent like dimethyl phthalate. The plan for incineration must be worked out in advance with the incinerator operator. Solvents that contain a high concentration of peroxides should be treated to remove peroxides before being sent to a secure landfill.

The greatest hazard is presented by a solid material that has crystallized from a peroxidizable solvent or that has been left after evaporation. If the material is in an open vessel it is feasible to add solvent *cautiously* to dissolve the peroxide as a dilute solution that can be treated to destroy the peroxide. However, if the material is in a long-unopened container, such as a screw-capped reagent bottle, there is a possibility that some peroxide may be present in the container closure and that it could be detonated by the friction of removing the closure. These materials should be handled and destroyed by people who are trained in dealing with explosives, such as a police bomb squad. Several serious accidents have resulted from the crystalline peroxide that is formed by diisopropyl ether on exposure to air.

Commonly used peroxide reagents, such as acetyl peroxide, benzoyl peroxide, *tert*-butyl hydroperoxide, and di-*tert*-butyl peroxide, are somewhat less treacherous than the adventitious peroxides formed in peroxidizable solvents, because their composition and properties are known and they are usually accompanied by information from the vendor or manufacturer on safe handling and disposal. The safe handling of such peroxides has been discussed by Varjavandi and Mageli.[25] These reagents are usually purchased or prepared in small quantities, and the even smaller unused excess can be disposed of by incineration, one container at a time; for safety reasons it is best first to dilute the peroxide with water or with a high-boiling solvent, such as a phthalate ester. These peroxides can also be burned in small batches in an isolated open ditch[25] if permitted by local regulations. The peroxide is spread thinly on the ground and ignited with a long-handled torch or an electrical squib. Peroxides with self-accelerating decomposition temperatures below room temperature (e.g., diisobutyryl peroxide, diisopropyl peroxydicarbonate) can be spread in the same kind of ditch and allowed to decompose for a day or so, provided that access to the disposal ditch is carefully controlled. To ensure complete destruction the residual decomposition products can then be burned. Organic peroxides (except for adventitious traces in solvents) are not permitted in secure landfills.

The sensitivity of most peroxides to shock and heat can be reduced by diluting them with inert solvents, and such dilution is recommended before transporting a peroxide for disposal. Alkanes are the most common choice, with phthalate esters or water as typical alternatives if different solvent characteristics are needed.

Because of the potential hazard in transporting peroxides and the small quantities generally at hand, they are candidates for decomposition by a chemist in the laboratory. Acidified ferrous sulfate solution is useful for destroying hydroperoxides, which are added to a solution containing about 50% molar excess ferrous sulfate with stirring at room temperature.

$$ROOH + 2Fe^{2+} + 2H^+ \longrightarrow ROH + 2Fe^{3+} + H_2O$$

Diacyl peroxides can be destroyed by this reagent as well as by aqueous sodium bisulfite, sodium hydroxide, or ammonia. However, diacyl peroxides with low solubility in water, such as dibenzoyl peroxide, react very slowly. A better reagent for destroying such peroxides is a solution of sodium or potassium iodide in glacial acetic acid.

$$(RCO_2)_2 + 2NaI \longrightarrow 2RCO_2Na + I_2$$

For 0.01 mol of diacyl peroxide, 0.022 mol (10% excess) of sodium or potassium iodide is dissolved in 70 mL of glacial acetic acid, and the peroxide is gradually added with stirring at room temperature. The solution is rapidly darkened by the formation of iodine. After a half hour the solution is washed down the drain with a large excess of water.

Most dialkyl peroxides (ROOR) do not react readily at room temperature with ferrous sulfate, iodide ion, ammonia, or the other reagents mentioned above. However, these peroxides can be destroyed by a modification of the iodide procedure.

One milliliter of 36% hydrochloric acid is added to the above acetic acid/potassium iodide solution as an accelerator. The dialkyl peroxide (0.01 mol) is added, and the solution is heated to 90–100°C on a steam bath over the course of a half hour and held at that temperature for 5 hours.

Q. DYES AND PIGMENTS

Dyes and organic pigments generally present no unusual problems for either incineration or burial in a landfill. Even though some dyes are sufficiently water soluble for drain disposal, this route should be limited to disposal of very small quantities so as not to impart color to the waste stream.

Arenediazonium salts ($ArN_2^+X^-$) are commonly used as intermediates for many kinds of organic synthesis. Some of them are unstable in solution, and many are explosive in the solid state.[26]

An arenediazonium salt can be disposed of by adding it gradually to a stirred solution of 5–10% excess 2-naphthol in 3% aqueous sodium hydroxide at 0–20°C. The resulting azo dye is separated by filtration and either incinerated or put into a secure landfill.

III. INORGANIC CHEMICALS

The majority of inorganic wastes can be considered to consist of a cationic part (metal or metalloid atoms) and an anionic part (often, but not always, nonmetallic atoms; many metals combine with nonmetallic atoms to form anions). In planning the disposal of these substances one should decide whether they present sufficiently low hazard from toxicity or other cause to be put into a sanitary landfill or the sewer system or whether the potential hazard is great enough to require disposal in a secure landfill. It is helpful to examine the cationic and anionic parts of the substance separately; if either part presents significant potential hazard, the substance should go to a secure landfill (possibly via an incinerator and the ash therefrom). Some of the elements themselves can present disposal problems because of reactivity (e.g., sodium) or toxicity (e.g., bromine).

Although it is sometimes assumed that if a substance contains a "heavy metal" it is highly toxic, this is not a sound basis for decision. While salts of some heavy metals, such as lead, thallium, and mercury, are highly toxic, those of others, such as gold and tantalum, are not. On the other hand, compounds of beryllium, a "light metal," are indeed highly toxic. In Table 6.1 cations of metals and metalloids are listed alphabetically in two groups: those whose toxic properties as described in toxicological literature[27] present a significant hazard and those whose properties do not. The basis for separation into two lists is relative and does not imply that those in the second list are "nontoxic": it is a basic precept of toxicology that everything, even water, is toxic under some conditions.

Similarly, Table 6.2 lists anions according to whether they present relatively high or low hazard. The distinctions in Table 6.2 are not based only on toxic properties; some of the groups possess other hazardous properties, such as strong oxidizing power (e.g., perchlorate), flammability (e.g., hydride), or explosivity (e.g., azide).

Elements that pose a hazard because of significant radioactivity are outside the scope of this report, and none is included in these two tables. Their handling and disposal are prescribed in detail in government regulations.[28]

Deciding on the disposal route for an inorganic waste with the aid of Tables 6.1 and 6.2 must be tempered with chemical judgment. There may be good reason for chemically altering a waste before disposal, as in the cases of reducing some strong oxidizing agents or converting toxic ions into insoluble salts.

A. Chemicals in Which neither the Cation nor the Anion Presents Significant Hazard

These are chemicals composed of ions from the right-hand columns of Tables 6.1 and 6.2. Chemicals of this type that are soluble in water to the extent of a few percent can usually be washed down the drain (see Chapter 5). Only laboratory quantities should be so disposed of, and at least 100 parts of water per part of chemical should be used. Local regulations should be checked for possible restrictions on certain ele-

TABLE 6.1   Relative Toxicity of Cations

High Toxic Hazard	Precipitant[a]	Low Toxic Hazard	Precipitant[a]
Antimony	OH^-, S^{2-}	Aluminum	OH^-
Arsenic	S^{2-}	Bismuth	OH^-, S^{2-}
Barium	SO_4^{2-}, CO_3^{2-}	Calcium	SO_4^{2-}, CO_3^{2-}
Beryllium	OH^-	Cerium	OH^-
Cadmium	OH^-, S^{2-}	Cesium	
Chromium (III)[b]	OH^-	Copper[c]	OH^-, S^{2-}
Cobalt (II)[b]	OH^-, S^{2-}	Gold	OH^-, S^{2-}
Gallium	OH^-	Iron[c]	OH^-, S^{2-}
Germanium	OH^-, S^{2-}	Lanthanides	OH^-
Hafnium	OH^-	Lithium	
Indium	OH^-, S^{2-}	Magnesium	OH^-
Iridium	OH^-, S^{2-}	Molybdenum (VI)[b,d]	
Lead	OH^-, S^{2-}	Niobium (V)	OH^-
Manganese (II)[b]	OH^-, S^{2-}	Palladium	OH^-, S^{2-}
Mercury	OH^-, S^{2-}	Potassium	
Nickel	OH^-, S^{2-}	Rubidium	
Osmium (IV)[b,e]	OH^-, S^{2-}	Scandium	OH^-
Platinum (II)[b]	OH^-, S^{2-}	Sodium	
Rhenium (VII)[b]	S^{2-}	Strontium	SO_4^{2-}, CO_3^{2-}
Rhodium (III)[b]	OH^-, S^{2-}	Tantalum	OH^-
Ruthenium (III)[b]	OH^-, S^{2-}	Tin	OH^-, S^{2-}
Selenium	S^{2-}	Titanium	OH^-
Silver	Cl^-, OH^-, S^{2-}	Yttrium	OH^-
Tellurium	S^{2-}	Zinc[c]	OH^-, S^{2-}
Thallium	OH^-, S^{2-}	Zirconium	OH^-
Tungsten (VI)[b,d]			
Vanadium	OH^-, S^{2-}		

[a]Precipitants are listed in order of preference:

   OH^- = base (sodium hydroxide or sodium carbonate)
   S^{2-} = sulfide
   Cl^- = chloride
   SO_4^{2-} = sulfate
   CO_3^{2-} = carbonate

[b]The precipitant is for the indicated valence state.
[c]Maximum tolerance levels have been set for these low-toxicity ions by the U.S. Public Health Service, and large amounts should not be put into public sewer systems. The small amounts typically used in laboratories will not normally affect water supplies.
[d]These ions are best precipitated as calcium molybdate or calcium tungstate.
[e]CAUTION: OsO_4, a volatile, extremely poisonous substance, is formed from almost any osmium compound under acid conditions in the presence of air.

TABLE 6.2 Relative Hazard of Anions

High-Hazard Anions

Ion	Hazard Type[a]	Precipitant	Low-Hazard Anions
Aluminum hydride, AlH_4	F	—	Bisulfite, HSO_3^-
Amide, NH_2^-	F, E[b]	—	Borate, BO_3^{3-}, $B_4O_7^{2-}$
Arsenate, AsO_3^-, AsO_4^{3-}	T	Cu^{2+}, Fe^{2+}	Bromide, Br^-
Arsenite, AsO_2^-, AsO_3^{3-}	T	Pb^{2+}	Carbonate, CO_3^{2-}
Azide, N_3^-	E, T	—	Chloride, Cl^-
Borohydride, BH_4^-	F	—	Cyanate, OCN^-
Bromate, BrO_3^-	O, E	—	Hydroxide, OH^-
Chlorate, ClO_3^-	O, E	—	Iodide, I^-
Chromate, CrO_4^{2-}, $Cr_2O_7^{2-}$	T, O	c	Oxide, O^{2-}
Cyanide, CN^-	T	—	Phosphate, PO_4^{3-}
Ferricyanide, $Fe(CN)_6^{3-}$	T	Fe^{2+}	Sulfate, SO_4^{2-}
Ferrocyanide, $Fe(CN)_6^{4-}$	T	Fe^{3+}	Sulfite, SO_3^{2-}
Fluoride, F^-	T	Ca^{2+}	Thiocyanate, SCN^-
Hydride, H^-	F	—	
Hydroperoxide, O_2H^-	O, E	—	
Hydrosulfide, SH^-	T	—	
Hypochlorite, OCl^-	O	—	
Iodate, IO_3^-	O, E	—	
Nitrate, NO_3^-	O	—	
Nitrite, NO_2^-	T, O	—	
Perchlorate, ClO_4^-	O, E	—	
Permanganate, MnO_4^-	T, O	d	
Peroxide, O_2^{2-}	O, E	—	
Persulfate, $S_2O_8^{2-}$	O	—	
Selenate, SeO_4^{2-}	T	Pb^{2+}	
Selenide, Se^{2-}	T	Cu^{2+}	
Sulfide, S^{2-}	T	e	

[a]Toxic, T; oxidant, O; flammable, F; explosive, E.
[b]Metal amides readily form explosive peroxides on exposure to air (see this chapter, Section III.C.7).
[c]Reduce and precipitate as Cr(III); see Table 6.1.
[d]Reduce and precipitate as Mn(II); see Table 6.1.
[e]See Table 6.3.

ments, such as copper. Dilute slurries of insoluble materials, such as calcium sulfate or aluminum oxide, also can be handled in this way, provided the material is finely divided and not contaminated with tar so that it is unlikely to clog the piping. Some incinerators can handle these chemicals.

When the above options are not available, the chemicals must go to landfill. For chemicals of this type a sanitary landfill can be used if local regulations permit. However, if the solution or suspension is dilute, the cost of landfilling, which is based on volume or gross weight, may make some kind of concentration attractive. If time and space requirements permit, dilute aqueous solutions can be boiled down or allowed to evaporate to leave only a sludge or the inorganic solid for landfill disposal. Alternatively, toxic ions can be concentrated into solid form by passing their solutions over ion-exchange resins. The resins can be landfilled, and the effluent solutions can be put down the drain.

Another procedure is to precipitate the metal ion by the agent recommended in Table 6.1 and send the precipitate to landfill. The most generally applicable procedure is to precipitate the cation as the hydroxide or oxide by adjusting the pH to the range shown in Appendix J. Because the pH range for precipitation varies greatly among metal ions, it is important to control pH carefully. The aqueous solution of the metal ion is adjusted to the recommended pH by addition of a 1 M solution of sulfuric acid or sodium hydroxide or carbonate. The pH can be determined over the range 1–10 by use of wide-range pH paper. For some ions the hydroxide precipitate will redissolve at a high pH. For a number of metal ions the use of sodium carbonate will result in precipitation of the metal carbonate or a mixture of hydroxide and carbonate, a matter of no importance.

The precipitate is separated by filtration, or as a heavy sludge by decantation, and packed for landfill disposal. Some gelatinous hydroxides are hard to filter. In such cases heating the mixture close to 100°C or stirring in an amount of diatomaceous earth estimated to be 1–2 times the weight of the precipitate often facilitates filtration. As shown in Table 6.1, precipitants other than base are superior for some metals, e.g., sulfuric acid for calcium.

B.   CHEMICALS IN WHICH A CATION PRESENTS A RELATIVELY HIGH HAZARD FROM TOXICITY

In general, waste chemicals containing any of the cations listed under High Toxic Hazard in Table 6.1 should be disposed of in a secure landfill. In many cases the wastes can be packaged in a lab pack without treatment. Economics dictates whether it is worth the trouble to convert an aqueous solution of a salt of a toxic cation into an insoluble material in order to reduce the volume of material in the lab pack. Bear in mind that explosive, pyrophoric, and reactive materials (except cyanides and sulfides) are not permitted in landfills (see Appendix A, 40 CFR 265.316).

The methods of concentrating or separating metal ions described above in Section III.A also apply here. Most of the high toxic hazard cations listed in Table 6.1 can be separated from aqueous solutions as the hydroxide or oxide (see Appendix J).

Alternatively, most of them can be precipitated as insoluble sulfides by treatment with sodium sulfide in neutral solution. Inasmuch as several of the sulfides will redissolve in excess sulfide ion, it is important that the sulfide ion concentration be controlled by adjustment of the pH.

For precipitation the pH of the metal ion solution is adjusted to neutral by addition of 1 M sodium hydroxide or 1 M sulfuric acid solution, using wide-range pH paper. A 1 M solution of sodium sulfide is added to the metal ion solution, and the pH is again adjusted to neutral with 1 M sulfuric acid solution. As with the hydroxides, the precipitate is separated by filtration or decantation and packed for landfill disposal. Excess sulfide ion can be destroyed by addition of hypochlorite to the clear aqueous phase (this chapter, Section III.C. 3). Guidance on precipitating many cations as sulfides is provided in Table 6.3.

The following ions are most commonly found as oxyanions and are not precipitated by base: As^{3+}, As^{5+}, Re^{7+}, Se^{4+}, Se^{6+}, Te^{4+}, and Te^{6+}. These elements can be precipitated from their oxyanions as the sulfides by the above procedure. Oxyanions of Mo^{6+} and W^{6+} can be precipitated as their calcium salts by addition of calcium chloride.

Another class of compounds whose cations may not be precipitated by addition of hydroxide ion are the more stable complexes of metal cations with such Lewis bases as ammonia, amines, and phosphites. Because of the large number of these compounds and their wide range of properties, it is not possible to give a general procedure for separating the cations. In many cases metal sulfides can be precipitated directly from aqueous solutions of the complexes by the addition of aqueous sodium sulfide. If a test-tube experiment shows that stronger measures are needed, the addition of hydrochloric acid to produce a slightly acidic solution will often decompose the complex by protonation of the basic ligand; metal ions that precipitate as sulfides under acid conditions can be subsequently precipitated by dropwise addition of aqueous sodium sulfide.

A third option for these wastes is incineration, provided that the incinerator ash is to be sent to a secure landfill. Incineration to ash reduces the volume of wastes going to a landfill. Wastes that contain mercury, thallium, gallium, osmium, selenium, or arsenic should not be incinerated because volatile, toxic combustion products may be emitted.

TABLE 6.3 Precipitation of Sulfides[a]

Precipitated at pH 7	Not Precipitated at Low pH	Forms a Soluble Complex at High pH
Ag^+		
As^{3+b}		X
Au^{+b}		X
Bi^{3+}		
Cd^{2+}		
Co^{2+}	X	
Cr^{3+b}		
Cu^{2+}		
Fe^{2+b}	X	
Ge^{2+}		X
Hg^{2+}		X
In^{3+}	X	
Ir^{4+}		X
Mn^{2+b}	X	
Mo^{3+}		X
Ni^{2+}	X	
Os^{4+}		
Pb^{2+}		
Pd^{2+b}		
Pt^{2+b}		X
Re^{4+}		
Rh^{2+b}		
Ru^{4+}		
Sb^{3+b}		X
Se^{2+}		X
Sn^{2+}		X
Te^{4+}		X
Tl^{+b}	X	
V^{4+b}		
Zn^{2+}	X	

[a]From Reference 29.
[b]Higher oxidation states of this ion are reduced by sulfide ion and precipitated as this sulfide.

C. CHEMICALS IN WHICH AN ANION PRESENTS A RELATIVELY HIGH HAZARD

The more common hazardous anions are listed in Table 6.2. Much that was said about the disposal of chemicals containing a hazardous cation applies equally to this class. Note that the hazard associated with some of these anions is explosivity or reactivity, which bars them from landfill disposal as such. Others can be put into lab packs for landfill disposal. Most chemicals containing these anions can be incinerated, but strong oxidizing agents and hydrides should be introduced into the incinerator in containers of not more than a few hundred grams, one at a time. Incinerator ash from anions of chromium or manganese should go to a secure landfill.

Some of these anions can be precipitated as insoluble salts for landfill disposal, as indicated in Table 6.2. Small amounts of strong oxidizing agents, hydrides, cyanides, azides, metal amides, and soluble sulfides and fluorides can be converted into less hazardous substances in the laboratory before being disposed of. Suggested procedures are presented in the following paragraphs.

1. Oxidizing Agents

Hypochlorites, chlorates, bromates, iodates, periodates, inorganic peroxides and hydroperoxides, persulfates, chromates, molybdates, and permanganates can be reduced by sodium bisulfite.

A dilute (< 5%) solution or suspension of a salt containing one of these anions is brought to pH < 3 with sulfuric acid, and a 50% excess of aqueous sodium bisulfite is added gradually with stirring at room temperature. The reaction mixture should contain a thermometer, and an increase in temperature indicates that reaction is taking place. If the reaction does not start on addition of about 10% of the sodium bisulfite, lowering the pH may initiate it. Colored anions (e.g., permanganate, chromate) serve as their own indicators of completion of the reduction. The reduced mixtures can usually be washed down the sink. However, if large amounts of permanganate or chromate have been reduced, it may be necessary to send the product to a secure landfill, possibly after reduction of volume by concentration or precipitation as described above in Section III.A.

Hydrogen peroxide can be reduced by the above bisulfite procedure or by ferrous sulfate as described for organic hydroperoxides (see Section II.P). However, it is usually acceptable to dilute it to a concentration less than 3% and flush it down the drain. Solutions with a hydrogen peroxide concentration greater than 30% should be handled with great care to avoid contact with reducing agents, including all organic materials, or with transition metal compounds, which can catalyze violent reactions.[30]

Concentrated perchloric acid (particularly >60%) must be kept away from reducing agents, including weak reducing agents such as ammonia, wood, paper, plastics, and all other organic substances, because it can react violently with them. Dilute perchloric acid is not reduced by common laboratory reducing agents such as bisulfite, hydrogen sulfide, hydriodic acid, iron, or zinc. Accordingly, perchloric acid is best disposed of by stirring it gradually into enough cold water to make its concentration less than 5%, neutralizing it with aqueous sodium hydroxide, and washing the solution down the drain with a large excess of water.[31]

Nitrate is most dangerous in the form of concentrated nitric acid (70% or higher), which is a potent oxidizing agent for organic material and all other reducing agents. It can also cause serious skin burns. Dilute aqueous nitric acid is not a dangerous oxidizing agent and is not easily reduced by common laboratory reducing agents. Nitric acid should be neutralized with aqueous sodium hydroxide before disposal down the drain; concentrated nitric acid should be diluted before neutralization. Metal nitrates are generally quite soluble in water. Those of the metals listed under low toxic hazard in Table 6.1, as well as ammonium nitrate, should be kept separate from oil or other organic materials because on heating such a combination a fire or explosion can occur.[32] Otherwise they can be treated as chemicals that present no significant hazard.

Nitrites in aqueous solution can be destroyed by adding about 50% excess aqueous ammonia and acidifying to pH 1 (use pH paper) with hydrochloric acid.

$$HNO_2 + NH_3 \xrightarrow{\ H^+\ } N_2 + 2H_2O$$

2. Metal Hydrides

Most metal hydrides react violently with water with evolution of hydrogen, which can form an explosive mixture with air. Some, such as NaH and LiAlH$_4$, are pyrophoric. Most can be decomposed by gradual addition of methanol, ethanol, *n*-butyl alcohol, or *tert*-butyl alcohol to a stirred, ice-cooled solution or suspension of the hydride in an inert liquid, such as diethyl ether, tetrahydrofuran, or toluene under nitrogen in a three-necked flask. Four common hydrides in laboratories are lithium aluminum hydride (LiAlH$_4$), sodium borohydride (NaBH$_4$), sodium hydride (NaH), and calcium hydride (CaH$_2$). The methods for their disposal that follow indicate that the reactivity of metal hydrides varies over a wide range. Most hydrides can be safely decomposed by one of the four methods, but the properties of a given hydride should be sufficiently well understood to select the most appropriate method.

Lithium aluminum hydride can be purchased as a solid or as 0.5–1.0 M solutions in toluene, diethyl ether, tetrahydrofuran, or other ethers. Although dropwise addition of water to its solutions under nitrogen in the apparatus described in the preceding paragraph has frequently been used to decompose it, vigorous frothing often occurs. An alternative is to use 95% ethanol, which reacts less vigorously than water. A safer procedure is to decompose the hydride with ethyl acetate because no hydrogen is formed[33]:

$$2CH_3CO_2C_2H_5 + LiAlH_4 \longrightarrow LiOC_2H_5 + Al(OC_2H_5)_3$$

The mixture sometimes becomes sufficiently viscous after the addition of ethyl acetate to make stirring difficult.[34] When reaction with ethyl acetate has ceased, a saturated aqueous solution of ammonium chloride is added with stirring. The mixture separates into an organic layer and an aqueous layer containing inert inorganic solids. The upper, organic layer should be packed for incineration or landfill disposal. The lower, aqueous layer can be flushed down the drain.

Sodium borohydride is so stable to water that a 12% aqueous solution stabilized with sodium hydroxide is an article of commerce. To decompose it the solid or aqueous solution is added to enough water to make the borohydride concentration less than 3%, and then excess dilute aqueous acetic acid is added dropwise with stirring under nitrogen.

Sodium hydride in the dry state is pyrophoric. It can be purchased as a relatively safe dispersion in mineral oil. Either form can be decomposed by adding enough dry hydrocarbon solvent (e.g., heptane) to reduce the hydride concentration below 5% and adding excess *tert*-butyl alcohol dropwise under nitrogen with stirring. Cold water is then added gradually, and the two layers are separated. The organic layer can be incinerated or sent to a secure landfill, and the aqueous layer can be flushed down the drain.

Calcium hydride is purchased as a powder, usually for drying ethers, esters, alcohols with more than three carbon atoms, and other solvents.[35] It is decomposed by adding 25 mL of methanol per gram of hydride under nitrogen with stirring. When the reaction has finished, an equal volume of water is gradually added to the slurry of calcium methoxide, and the mixture is then flushed down the drain with a large volume of water.

3. Inorganic Sulfides

Small amounts of unused sodium or potassium sulfide reagent can be destroyed in aqueous solution by sodium or calcium hypochlorite, using the procedure described for oxidizing mercaptans (this chapter, Section II.F).

$$Na_2S + 4OCl^- \longrightarrow Na_2SO_4 + 4Cl^-$$

Small amounts of hydrogen sulfide can be oxidized by the same reagents. To keep the reaction under control, one first absorbs the hydrogen sulfide in excess aqueous sodium hydroxide and then adds the solution dropwise to the hypochlorite solution in a good hood. (CAUTION: Hydrogen sulfide is an acute poison and is one of the most dangerous chemicals in the laboratory.[36])

An alternative is to precipitate the sulfide ion as an insoluble sulfide (Table 6.3) for landfill disposal.

4. Inorganic Fluorides

One should be familiar with the properties and proper handling of hydrogen fluoride (HF) before working with it.[37] Hydrogen fluoride can cause serious burns, with the risk from aqueous solutions increasing with concentration. Waste aqueous hydrofluoric acid can be added slowly to a stirred solution of excess slaked lime to precipitate calcium fluoride, which is chemically inert and has little or no toxic hazard. If no more than about 100 g of calcium fluoride is formed, the suspension can be washed down the drain; larger amounts should be separated and sent to a landfill.

As an alternative, hydrofluoric acid can be diluted to about 1% concentration by adding it with stirring to cold water in a polyethylene or borosilicate glass vessel, neutralizing the acid with aqueous sodium hydroxide, and precipitating calcium fluoride by addition of excess calcium chloride solution.

Soluble metal fluorides such as sodium and potassium fluorides are highly poisonous. They can be converted into calcium fluoride by treating aqueous solutions with calcium chloride solution. Boron trifluoride can be dissolved in water and the fluoride ion precipitated as calcium fluoride by adding calcium chloride solution. The calcium fluoride is filtered and sent to a landfill, and the filtrate is flushed down the drain.

$$2BF_3 + 3CaCl_2 + 6H_2O \longrightarrow 3CaF_2 + 2B(OH)_3 + 6HCl$$

5. Inorganic Cyanides

Hydrogen cyanide, b.p. 26°C, is among the most lethal chemicals known and is toxic by inhalation, ingestion, or skin absorption. Precautions for the safe handling of hydrogen cyanide are described in the literature.[38] It can be disposed of by dilution with one or more volumes of ethanol followed by incineration of the solution. Small amounts can be oxidized to relatively innocuous cyanate by aqueous sodium hypochlorite, using the procedure described for oxidizing mercaptans (this chapter, Section II.F). The oxidation should be carried out in a good hood using adequate precautions.[38]

$$NaCN + NaOCl \longrightarrow NaOCN + NaCl$$

The hydrogen cyanide is dissolved in several volumes of ice water in an ice-cooled, three-necked flask equipped with a stirrer, thermometer, and dropping funnel. Approximately one molar equivalent of aqueous sodium hydroxide is added at 0–10°C to convert the hydrogen cyanide into its sodium salt. (CAUTION: Sodium hydroxide or other bases, including sodium cyanide, must not be allowed to come into contact with liquid hydrogen cyanide because they may initiate a violent polymerization of hydrogen cyanide.) A 50% excess of commercial laundry bleach (e.g., Clorox) containing 5.25% (0.75 molar) sodium hypochlorite is added at 0–10°C. When the addition is complete and heat is no longer being evolved, the solution is allowed to warm to room temperature and stand for several hours. The mixture is washed down the drain with excess water.

The same procedure can be applied to soluble cyanides, such as sodium cyanide and potassium cyanide, as well as to insoluble cyanides such as cuprous cyanide. The procedure also destroys soluble ferrocyanides and ferricyanides; alternatively, these can be precipitated as the ferric or ferrous salt, respectively, for landfill disposal. In calculating the quantity of hypochlorite required, remember that additional equivalents may be needed if the metal ion can be oxidized to a higher valence state, as in the reaction:

$$2CuCN + 3NaOCl + H_2O \longrightarrow$$
$$2Cu^{2+} + 3Na^+ + 2OCN^- + 2OH^- + 3Cl^-$$

6. Metal Azides

Heavy-metal azides are notoriously explosive. Sodium azide, a common preservative in clinical laboratories and a useful reagent in synthetic work, is not explosive except when heated near its decomposition temperature (300°C); heating sodium azide should be avoided. Sodium azide should never be flushed down the drain; this practice has caused serious incidents because the azide has reacted with lead or copper in the drain lines to produce azides that have later exploded.[39] Moreover, sodium azide has high acute toxicity as well as high toxicity to bacteria in water-treatment plants. It can be destroyed by reaction with nitrous acid.[39]

$$2NaN_3 + 2HNO_2 \longrightarrow 3N_2 + 2NO + 2NaOH$$

The operation must be carried out in a hood because of the formation of nitric oxide. An aqueous solution containing no more than 5% sodium azide is put into a three-necked flask equipped with a stirrer, a dropping funnel, and an outlet with plastic tubing to carry nitrogen oxides to the hood flue. A 20% aqueous solution of sodium nitrite containing 1.5 g (about 40% excess) of sodium nitrite per gram of sodium azide is added with stirring. A 20% aqueous solution of sulfuric acid is then added gradually until the reaction mixture is acidic to litmus paper. (CAUTION: This order of addition is essential. If the acid is added before the nitrite, poisonous, volatile HN_3 will be evolved.) When the evolution of nitrogen oxides ceases, the acidic solution is tested with starch-iodide paper; if it turns blue, excess nitrite is present and decomposition is complete. The reaction mixture is washed down the drain.

Wear[39] also describes ways to decompose lead azide and to decontaminate drain lines suspected of containing lead azide or copper azide, using nitrous acid or ceric ammonium nitrate. However, such potentially dangerous operations should be carried out only by people with experience in handling explosives and never by laboratory workers.

Other alkali metal azides can be treated like sodium azide. Hydrazoic acid (HN_3), besides being highly toxic, is highly explosive and should never be isolated. Solutions of hydrazoic acid in water, benzene, or chloroform are quite stable.

Solutions in benzene or chloroform can be decomposed by adding enough water to make the concentration of hydrazoic acid in the water less than 5% and stirring vigorously or shaking in a separatory funnel to extract the HN_3 into the aqueous layer. The aqueous solution is neutralized with aqueous sodium hydroxide, separated from the organic layer, and treated by the procedure for decomposing sodium azide.

7. Metal Amides

Bretherick[40] states that "many metal derivatives of nitrogenous systems linking nitrogen to a metal (usually but not exclusively a heavy metal) show explosive instability" and lists 81 examples, including Ag_2NH, Cu_3N, $Cd(NH_2)_2$, and PbNH. It is difficult to give general directions for disposing of this large, diverse group, most of them encountered only rarely, except to say that only workers experienced in handling explosive substances should be working with most of them at all. Those that are not inherently explosive can be incinerated, dropping in one bottle at a time.

Two of the compounds listed by Bretherick, sodium amide and potassium amide, are common reagents that are relatively safe to use and dispose of if proper precautions are taken. They may ignite or explode on heating or grinding in air, particularly if they have been previously exposed to air or moisture, because of explosive oxidation products that may have formed on the surface.[41] The presence of such oxidation products is signaled by a tan, brown, or orange color, contrasted with the normal white or light gray color of the unoxidized amide. Any such discolored amide should be decomposed by the hydrocarbon-ethanol procedure given below. The explosivity problem can be avoided by preparing these amides in the vessel in which they will be used just before they are needed.[42] Sodium amide free of explosive oxides can be purchased in sealed bottles. The bottles should be opened only once, to take out the quantity of reagent needed, and the remainder of the bottle contents should be disposed of promptly. It has been recommended that in the laboratory sodium amide be decomposed by covering it with toluene or kerosene and adding 95% ethanol slowly with stirring.[43] Alternatively, sodium amide in small portions can be stirred into excess solid ammonium chloride.[44]

$$NaNH_2 + NH_4Cl \longrightarrow NaCl + 2NH_3$$

These procedures are also applicable to potassium amide.

D. METALS

1. Alkali Metals

Alkali metals react violently with water, with common hydroxylic solvents, and with halogenated hydrocarbons. Potassium can form explosive peroxides on exposure to air. Because of these properties, alkali metals cannot be put into landfills. Bottles containing up to about 100 g can be incinerated.

Waste sodium is readily destroyed with 95% ethanol using a modification of a procedure described in the literature.[45]

The procedure is carried out in a three-necked, round-bottomed flask equipped with a stirrer, dropping funnel, water-cooled condenser, and heating mantle or steam bath. Solid sodium should be cut into small pieces while wet with a hydrocarbon, preferably mineral oil, so that the unoxidized surface is exposed; a dispersion of sodium in mineral oil can be treated directly. The flask is flushed with nitrogen and the sodium put into it. Then 13 mL of 95% ethanol per gram of sodium is added at a rate that causes rapid refluxing. Stirring is started as soon as enough ethanol has been added to make it feasible. The mixture is stirred and heated under reflux until the sodium is dissolved. Heat is turned off, and an equal volume of water is added at a rate that causes no more than mild refluxing. The solution is then cooled, neutralized with 6 N sulfuric or hydrochloric acid, and washed down the drain.

Lithium metal can be treated by the same procedure, but using 30 mL of 95% ethanol per gram of lithium. The rate of solution is lower than that of sodium.[46]

To destroy the more reactive potassium metal, the less reactive *tert*-butyl alcohol is used in the proportion of 21 mL per gram of metal.[47] If the potassium is dissolving too slowly, a few percent of methanol can be added gradually to the refluxing *tert*-butyl alcohol. (CAUTION: Potassium metal that has formed a yellow oxide coating from exposure to air should not be cut with a knife, even when wet with a hydrocarbon, because of the explosive property of the oxide.[48]) Oxide-coated potassium sticks should be put directly into the flask and decomposed with *tert*-butyl alcohol; the decomposition will require considerable time because of the low surface/volume ratio of the metal sticks.

2. Other Metals

Most metals are pyrophoric if sufficiently finely divided.[49] Above some critical ratio of surface area to mass that is specific for each metal, the heat of oxide formation on exposure to air can lead to ignition or to a dust explosion. Among the metals best known for this type of pyrophoricity are aluminum, cobalt, iron, magnesium, manganese, nickel, palladium, platinum, titanium, tin, uranium, zinc, zirconium, and their alloys. Any of these metals in finely divided form should be kept under an inert atmosphere in a tightly closed container. During handling these metal powders should be exposed to air as little as possible; a fire extinguisher that has a special granular formulation designed to control burning metal (e.g., Met-L-X extinguishers) should be close at hand.

These metal powders can be disposed of by adding enough water to make a paste, which is then spread a few millimeters thick on a metal pan to dry in the open air. As the paste dries over the course of a day or two, the surface of the metal particles gradually oxidizes; the resulting metal is no longer pyrophoric and can be transferred to a container for disposal in a landfill.

Nickel, palladium, platinum, and other metals or combinations of them that are used as hydrogenation catalysts should never be allowed to dry in air following a hydrogenation, for example on filter paper, because they are prone to ignite. Filter cakes containing such catalysts should be promptly moistened with water. Used nickel catalyst can be dissolved in hydrochloric acid as described earlier in this chapter in Section II.O. Used palladium and platinum catalysts should be handled according to the procedures in Chapter 4.

E. WATER-REACTIVE METAL HALIDES

These compounds cannot be put into a landfill because of the hazard characteristic of reactivity (see Appendix A, 40 CFR 265.316). Liquid halides, such as $TiCl_4$ and $SnCl_4$, can be added to well-stirred water in a round-bottomed flask cooled by an ice bath as necessary to keep the exothermic reaction under control. It is usually more convenient to add solid halides, such as $AlCl_3$ and $ZrCl_4$, to stirred water and crushed ice in a beaker. The metal ions can be separated and disposed of as described above in Section III.A.

F. HALIDES AND ACID HALIDES OF NONMETALS

Examples of this class are BCl_3, PCl_3, $SiCl_4$, $SOCl_2$, SO_2Cl_2, and $POCl_3$. Most of them are water reactive and cannot be landfilled (see Appendix G). The liquids can be conveniently hydrolyzed with 2.5 N sodium hydroxide by the procedure given earlier in this chapter in Section II.J, with the hydrolysate flushed down the drain after neutralization. These compounds are irritating to the skin and respiratory passages and, even more than most chemicals, require good hoods and skin protection in handling them. Moreover, PCl_3 may give off small amounts of highly toxic phosphine (PH_3) during hydrolysis.

Sulfur monochloride, S_2Cl_2, is a special case. It is hydrolyzed to a mixture of sodium sulfide and sodium sulfite, so the hydrolysate must be treated with hypochlorite as described for metal sulfides (see Section III.C.3) before it can be flushed down the drain.

Because the solids of this class (e.g., PCl_5) tend to cake and fume in moist air, they are not conveniently hydrolyzed in a three-necked flask. It is better to add them to a 50% excess of sodium hydroxide in a beaker equipped with a stirrer and half-filled with crushed ice. If the solid has not all dissolved by the time the ice has melted and the stirred mixture has reached room temperature, the reaction can be completed by heating the stirred mixture on a steam bath.

Boron trifluoride can be converted into calcium fluoride (see Section III.C.4).

G. Nonmetal Hydrides

Nonmetal hydrides, such as B_2H_6, PH_3, and AsH_3, are sensitive to oxidation by many oxidizing agents, and many are pyrophoric (see Appendix H). They can be oxidized to safer materials by aqueous copper sulfate as described for white phosphorus. To avoid violent reaction with air, the oxidation must be carried out under nitrogen, most conveniently in a three-necked flask equipped with stirrer, nitrogen inlet, and dropping funnel or, if the hydride is gaseous, a gas inlet. The reaction mixture is worked up as described for white phosphorus.

H. Phosphorus

White phosphorus reacts with a number of oxidizing agents, including air, in which it is spontaneously flammable (see Appendix H). It is usually sold in the form of pellets weighing a few grams and is stored under water to prevent oxidation. If local regulations permit, gram quantities can be disposed of by being put in a pit in an open field and allowed to burn. Larger quantities can be incinerated by introducing small bottles of phosphorus covered with water into an incinerator one at a time.

In the laboratory, gram quantities of phosphorus can be oxidized by 1 M aqueous copper(II) sulfate, using twice as much reagent as called for by the equation

$$2P + 5Cu^{2+} + 8H_2O \longrightarrow 2H_3PO_4 + 5Cu + 10H^+$$

The reaction is more complex than shown; the phosphorus becomes coated with black copper phosphide that gradually forms crystalline copper and soluble phosphorous and phosphoric acids.[50]

Procedure for Oxidizing Five Grams of White Phosphorus. Five grams (0.16 mol) of white phosphorus is cut under water into pellets up to 5 mm across. The pellets are added to 800 mL (0.80 mol) of

1 M cupric sulfate solution in a 2-L beaker in a hood. The mixture is allowed to stand for about a week with occasional stirring. The phosphorus gradually disappears, and a fine black precipitate of copper and copper phosphide is formed. The reaction is complete when no waxy white phosphorus is observed when one of the larger black pellets is cut under water. The precipitate is separated on a Büchner funnel in the hood and, while still wet, transferred to 500 mL of laundry bleach (5.25% NaOCl) and stirred for an hour to ensure oxidation to phosphate of any copper phosphide, a potential source of toxic phosphine gas. The copper salt solution is treated as described above in Section III.A.

Red phosphorus is not pyrophoric and is much safer to work with than white phosphorus. Nevertheless, it is flammable and can create an explosion if mixed with a strong oxidant such as potassium permanganate. Red phosphorus can be disposed of by oxidation to phosphoric acid by aqueous potassium chlorate.

Procedure for Oxidizing Five Grams of Red Phosphorus.

$$6P + 5KClO_3 + 9H_2O \longrightarrow 6H_3PO_4 + 5KCl$$

Five grams (0.16 mol) of red phosphorus is added to a solution of 33 g (0.27 mol, 100% excess) of potassium chlorate in 2 L of 1 N sulfuric acid. The mixture is heated under reflux until the phosphorus has dissolved (usually 5–10 hours, depending on the particle size of the phosphorus). The solution is cooled to room temperature, treated with about 14 g of sodium bisulfite to reduce the excess chlorate (see Section III.C.1) and washed down the drain.

I. PHOSPHORUS PENTOXIDE

This water-reactive substance can be disposed of by gradually adding it to a mixture of water and crushed ice that half fills a beaker equipped with a stirrer. If there are still globs of unreacted pentoxide present when the mixture has reached room temperature, the mixture should be stirred and heated on a steam bath until all is dissolved. The solution is neutralized and flushed down the drain.

J. HYDRAZINE AND SUBSTITUTED HYDRAZINES

Hydrazine is a common reagent that is a carcinogen in animal tests and a strong skin irritant. It can be destroyed by a 25% excess of hypochlorite, following the procedure used for oxidizing mercaptans (this chapter, Section II.F). The hydrazine should be diluted to a concentration of about 5% with water before being added to ensure that the reaction will not be too vigorous.

$$N_2H_4 + 2OCl^- \longrightarrow N_2 + 2Cl^- + 2H_2O$$

Monosubstituted hydrazines, such as phenylhydrazine and methylhydrazine, can be decomposed in similar fashion.

$$RNHNH_2 + OCl^- \longrightarrow RH + N_2 + Cl^- + H_2O$$

Hypochlorite is also effective for destroying 1,1-dimethylhydrazine.[51]

IV. ORGANO-INORGANIC CHEMICALS

This class includes organometallics, i.e., compounds with metal-carbon bonds such as alkyllithiums, alkylaluminums, cyclopentadienyliron (ferrocene), and metal carbonyls. Many of these are sensitive to air and/or water. The class also includes compounds with metal atoms bound to organic molecules through oxygen, nitrogen, or other nonmetals; most of these compounds are fairly insensitive to air and water. Examples are metal alkanecarboxylates, metal alkoxides, metal acetylacetonates, and metal phthalocyanines.

Organo-inorganic chemicals that are insensitive to air or water can be disposed of by incineration or landfilling. If the metal involved is relatively toxic (see Table 6.1), the compound or its ash must go to a secure landfill. Appendixes G and H list the more common kinds of organo-inorganic compounds that are water reactive or pyrophoric, respectively, and information on their disposal in the laboratory is given below.

A. ORGANOMETALLICS SENSITIVE TO AIR OR WATER

Common examples of this group are Grignard reagents, alkyllithiums or aryllithiums, and trialkylaluminums, which are prepared, purchased, or used as dilute solutions in an ether or hydrocarbon. Less common examples are arylsodiums and dialkylzincs. The trialkylaluminums and dialkylzincs are often prepared and stored solvent-free; the first step in decomposing such compounds is to dissolve them under nitrogen in about 20 volumes of dry toluene, heptane, or some other hydrocarbon to make them less likely to ignite and easier to handle.

A solution of the organometallic (less than 5% concentration) in a hydrocarbon or an ether is placed in a three-necked flask equipped with a stirrer, dropping funnel, nitrogen inlet, and ice bath. A 10% excess of *tert*-butyl alcohol dissolved in a hydrocarbon is added dropwise to the well-stirred solution under nitrogen. This is followed by addition of cold water and then enough 5% hydrochloric acid to neutralize the aqueous phase. The aqueous and organic phases are separated and disposed of appropriately. Where necessary, as for example with a dialkylcadmium, metal ions can be separated from the aqueous phase by procedures given above in Section III.

B. METAL CARBONYLS

Metal carbonyls are highly toxic materials, and many are sensitive to air. Because the metals are generally in a low oxidation state, the carbonyls can be destroyed by oxidation, in most cases by the gradual addition of the carbonyl to 25% excess of a well-stirred hypochlorite solution. The equipment and procedure used in the hypochlorite oxidation of mercaptans (this chapter, Section II.F) are appropriate, with the proviso that the reaction should be carried out under nitrogen and the metal carbonyl should be added as an approximately 5% solution in an inert solvent such as tetrahydrofuran or a hydrocarbon.

$$Ni(CO)_4 + OCl^- + H_2O \longrightarrow Ni^{2+} + Cl^- + 2OH^- + 4CO$$

C. NONMETAL ALKYLS AND ARYLS

Nonmetal alkyls and aryls, such as BR_3, PR_3, and AsR_3, are sensitive to oxidation by air and other oxidizing agents, although aryls are more stable than the alkyls. Many of these compounds are highly toxic. As with organic sulfides (this chapter, Section II.G), most of these compounds can be oxidized by 25% excess sodium hypochlorite (laundry bleach) to the corresponding oxides, which, since they are not pyrophoric, can be landfilled.

The alkyl nonmetal hydrides, such as R_2PH and $RAsH_2$, can generally be decomposed by this hypochlorite procedure. However, a few are so readily oxidized that they react violently with air, and for them the copper sulfate oxidation procedure described for nonmetal hydrides (Section III.G above) should be used.

D. ORGANOMERCURY COMPOUNDS

Organomercury compounds, such as R_2Hg and $RHgX$, should not be incinerated because some mercury may be volatilized. In many cases they can be oxidized as a dispersion in 25% excess laundry bleach (5.25% sodium hypochlorite).

$$R_2Hg + 2OCl^- + H_2O \longrightarrow HgO + 2ROH + 2Cl^-$$

Because of water insolubility, this reaction may be slow. In such a case, addition of a solution of bromine in carbon tetrachloride to a solution or suspension of the organomercury compound in carbon tetrachloride will result in oxidation to mercuric bromide, which can be separated and sent to a landfill.

E.   CALCIUM CARBIDE

This reagent is no longer common in the laboratory but is occasionally used as a source of acetylene.[52] Small amounts of water in contact with calcium carbide can generate explosive acetylene/air mixtures that can be ignited by locally overheated carbide. The following procedure for decomposing small amounts of calcium carbide with aqueous acid minimizes this hazard.

$$CaC_2 + 2HCl \longrightarrow CaCl_2 + C_2H_2$$

Fifty grams of calcium carbide is suspended in 600 mL of a hydrocarbon such as toluene or cyclohexane in a 2-L three-necked flask equipped with an ice bath, stirrer, dropping funnel, nitrogen inlet, and a gas outlet leading through plastic tubing to the back of the hood that the equipment is in. With a moderate flow of nitrogen passing through the flask to carry off the acetylene generated, 300 mL of 6 N hydrochloric acid is added dropwise over a period of about 5 hours, and the mixture is stirred an additional hour. The aqueous and hydrocarbon layers are separated and disposed of separately; the aqueous layer can be flushed down the drain after being neutralized.

V.   CLASSIFICATION OF UNLABELED CHEMICALS FOR DISPOSAL

A laboratory chemical in a container with a missing label must be characterized as to general type (e.g., aromatic hydrocarbon, chlorinated solvent, mineral acid, soluble metal salt) so that it can be disposed of safely. It will generally suffice to carry out the preliminary steps of the well-known *Systematic Identification of Organic Compounds*.[53] These steps are outlined below; the significance of the observations is discussed in the reference. Although the procedure is designed for organic substances, most of the steps and the same principles also apply to inorganic substances. These steps can also be used to classify, for disposal purposes, an orphan reaction mixture—an unlabeled material in laboratory glassware that has been left behind by a departed laboratory worker.

If the chemical is in a reagent container, examine the container for any clues as to the supplier, which may be revealing. In the case of orphan reaction mixtures, knowledge of the area of chemistry in which the departed worker was working may provide useful clues as to chemical type. The size of the container may be informative; a large one is likely to hold a common chemical. Observe whether the substance is a solid or a liquid, whether it is fluid or viscous. See if it is acidic or basic to pH paper. Cautiously sniff the odor of the cap.

Put about 0.1 g of the substance in a small porcelain dish or a metal spoon. Bring a flame near to see if the substance burns and, if so, how readily and with what appearance (for example, a sooty yellow flame suggests an aromatic ring compound). Heat it gently, then strongly; this gives an indication of volatility. If the substance is a solid, does it melt or decompose? Is there a residue? If so, it is probably an inorganic salt or oxide; add a drop of water to see whether it dissolves and, if so, whether the solution is acidic or basic to pH paper.

Pick up a little of the substance in a small loop in a copper wire and hold in a flame; halogenated compounds impart a distinct green color to the flame.

Test the solubility of the substance in water and, as seems called for, in ether, dilute sodium hydroxide, dilute hydrochloric acid, and concentrated sulfuric acid; the "solubility class"[53] is very informative. If it is insoluble in any solvent tested, note whether it floats or sinks—this gives an idea of its density.

These steps can be carried out in a few minutes by an experienced chemist and often provide an adequate classification of the unknown chemical without further investigation. If more information is needed, the next step can be a simple infrared scan, which can indicate functional groups and whether the substance is a single entity or a complex mixture. Additional testing can involve some of the simpler "classification tests,"[53] such as testing for oxidizability with dilute aqueous potassium permanganate or for loosely bound halogen with alcoholic silver nitrate.

Some of the common inorganic oxidizing agents have characteristic colors, e.g., chromates are yellow, permanganates are purple. Starch iodide paper or other test paper (this chapter, Section II.P) can be used to indicate organic or inorganic oxidizing agents.

If the chemical is inorganic and is not pyrophoric, water reactive, or a strong oxidizing agent, it is generally not necessary to identify it precisely or to treat it before disposal in a secure landfill. Should it be necessary to identify the cation or anion, e.g., for toxic characteristics, good procedures are provided by Sorum.[54]

The results of the kinds of tests outlined above will generally provide sufficient information about an unknown chemical for assignment of its hazard class and selection of a safe method of disposal.

VI. DISPOSAL OF LEAKING OR UNIDENTIFIED GAS CYLINDERS

An earlier report of this committee discusses the handling of leaking gas cylinders[55] as well as procedures for working with compressed gases.[56] As described in these references, if the leak cannot be remedied by tightening a valve gland or packing nut, the cylinder must be taken to an isolated, well-ventilated area. Then the cylinder may be vented cautiously if the gas is flammable or the gas slowly directed into an appropriate chemical neutralizer if it is corrosive or toxic.

Cylinders with unknown contents pose particularly difficult problems and considerable hazard. If the supplier can be identified, a phone call with a description of the cylinder may serve to identify the gas. If that fails, it is usually necessary to call in a waste-disposal firm that has the capability for dealing with the problem.

REFERENCES

1. L. F. Fieser and M. Fieser, *Reagents for Organic Synthesis*, John Wiley & Sons, New York, 10 vols. (1967–1982).
2. *Safety in the Chemical Laboratory*, Division of Chemical Education, American Chemical Society, Easton, Pa. (4 vols.).
3. NRC Committee on Hazardous Substances in the Laboratory, *Prudent Practices for Handling Hazardous Chemicals in Laboratories*, National Academy Press, Washington, D.C., 1981.
4. L. Bretherick, *Handbook of Reactive Chemical Hazards*, 2nd ed., Butterworths, London–Boston, 1979.
5. *Organic Syntheses*, John Wiley & Sons, New York.
6. *Organic Reactions*, John Wiley & Sons, New York.
7. IARC Scientific Publication No. 30, *Laboratory Decontamination and Destruction of Aflatoxins B_1, B_2, G_1, G_2 in Laboratory Wastes*, International Agency for Research on Cancer, Lyon, France, 1980. Aflatoxins in solution are destroyed by a large excess of hypochlorite solution or by acidified permanganate solution; on glassware by hypochlorite solution; and in animal feed by ammonia and heat.
8. IARC Scientific Publication No. 43, *Laboratory Decontamination and Destruction of Carcinogens in Laboratory Wastes: Some Nitrosamines*, International Agency for Research on Cancer, Lyon, France, 1982.
9. W. G. Mikell and L. R. Hobbs, *J. Chem. Educ.* 58, A165 (1981).
10. M. A. Armour, L. M. Browne, and G. L. Weir, *Hazardous Chemicals Information*

and Disposal Guide, Department of Chemistry, University of Alberta, Edmonton, Alta. T6G 2G2, Canada.

11. IARC Scientific Publication No. 33, *Handling Chemical Carcinogens in the Laboratory*, International Agency for Research on Cancer, Lyon, France, 1979, pp. 16–19.

12. Reference 1, Vol. 1, p. 935.

13. E. J. Keating, R. A. Brown, and E. S. Greenberg, *Ind. Water Eng.*, *15*, 22 (Dec. 1978).

14. R. L. Shriner and E. C. Kleiderer, *Org. Synth. Coll. Vol. 2*, 538 (1943); J. R. Ruhoff, *ibid.*, 315.

15. M. S. Newman and H. L. Holmes, *Org. Synth. Coll. Vol. 2*, 428 (1943).

16. N. Kornblum, *Org. React. 2*, 262 (1944).

17. Reference 4, pp. 161–164.

18. Reference 1, Vol. 1, p. 191.

19. H. L. Jackson, W. B. McCormack, C. S. Rondestvedt, K. C. Smeltz, and I. E. Viele, *Safety in the Chemical Laboratory 3*, 114–117 (1974).

20. D. R. Burfield, *J. Org. Chem. 47*, 3281 (1982).

21. Reference 3, pp. 63–68.

22. R. D. Mair and A. J. Graupner, *Anal. Chem. 36*, 194 (1964).

23. EM-Quant TM, MC/B Manufacturing Chemists, Inc., 2909 Highland Ave., Cincinnati, Ohio 45212.

24. W. Dasler and C. D. Bauer, *Ind. Eng. Chem. Anal. Ed. 18*, 52 (1946).

25. J. Varjavandi and O. L. Mageli, *Safety in the Chemical Laboratory 3*, 118–122 (1974).

26. Reference 4, pp. 46–50.

27. F. A. Patty, *Industrial Hygiene and Toxicology*, John Wiley & Sons, Interscience, New York, 1981, Vol. IIA, pp. 1493–2060; J. Doull, C. D. Klaasen, and M. O. Amdur, *Toxicology*, Macmillan, New York, 1980, pp. 409–467; *Documentation of the Threshold Limit Values for Substances in Workroom Air*, American Conference of Governmental and Industrial Hygienists, Cincinnati, Ohio, 1980.

28. U.S. Nuclear Regulatory Commission, *Rules and Regulations*, 10 CFR Chapter 1.

29. E. H. Swift and W. P. Schaefer, *Qualitative Elemental Analysis*, Freeman, San Francisco, 1961.

30. Reference 4, p. 959.

31. E. W. Bowerman, P. M. Harris, A. J. Krubsack, E. P. Schram, and T. R. Swett, *Safety in the Chemical Laboratory 2*, 97 (1971).

32. Reference 4, pp. 108, 122, and 994.

33. Reference 1, p. 583.

34. W. G. Brown, *Org. React. 6*, 490 (1951).

35. Reference 1, Vol. 1, p. 105.

36. Reference 3, pp. 99, 135–136.

37. Reference 3, pp. 43–45.

38. Reference 3, pp. 45–47, 133–135.

39. J. O. Wear, *Safety in the Chemical Laboratory 4*, 77 (1981).

40. Reference 4, p. 102.

41. Reference 4, pp. 955–957.

42. C. A. Vanderwerf and L. V. Lemmerman, *Org. Synth. Coll. Vol. 3*, 44 (1955); C. R. Hauser, J. T. Adams, and R. Levine, *ibid.*, p. 291.

43. F. W. Bergstrom, *Org. Synth. Coll. Vol. 3*, 782 (1955).

44. Reference 1, Vol. 1, p. 1034.

45. Reference 1, Vol. 1, p. 1065.

46. W. M. Jones, M. J. Grasley, and W. S. Brey, *J. Am. Chem. Soc. 85*, 2754 (1963).

47. W. S. Johnson and W. P. Schneider, *Org. Synth. Coll. Vol. 4*, 132 (1963).
48. Reference 4, p. 1025.
49. Reference 4, pp. 103, 119, and 170.
50. J. R. Van Wazer, *Phosphorus and Its Compounds*, Interscience Publishers, New York, (1958), Vol. 1, p. 112; H. Pelabon, *Bull. Soc. Chim. Fr.*, *53*, 260 (1933).
51. Product Bulletin, *Unsymmetrical Dimethylhydrazine*, Olin Chemicals, Stamford, Conn., 1981.
52. G. F. Woods and L. H. Schwartzman, *Org. Synth. Coll. Vol. 4*, 471 (1963).
53. R. L. Shriner, R. C. Fuson, D. Y. Curtin, and T. C. Morrill, *The Systematic Identification of Organic Compounds,* 6th ed., John Wiley & Sons, New York, 1980.
54. C. H. Sorum, *Introduction to Semimicro Qualitative Analysis*, Prentice-Hall, Englewood Cliffs, N.J., 1960, pp. 175–208.
55. Reference 3, pp. 236–237.
56. Reference 3, pp. 75–101, 221–224.

7 Disposal of Explosives from Laboratories

I. INTRODUCTION

An explosive chemical or mixture of chemicals is one that can undergo violent or explosive decomposition under appropriate conditions of reaction or initiation.[1] Some chemicals used in laboratories are known to be explosive, and some combinations of common laboratory reagents can be explosive (see Appendix F). Laboratory manipulations with known explosive chemicals or reagent combinations should be carried out only by personnel who are thoroughly familiar with the hazards involved, the precautions to be taken,[2] and procedures for destroying or disposing of potentially explosive materials. Any laboratory procedure that results in an unexpected explosion should be investigated to ascertain the probable cause and a laboratory safety rule established to prevent a recurrence. It is recommended that the circumstances of an unexpected explosion be submitted for publication in *Chemical and Engineering News*.

II. HANDLING EXPLOSIVES

A cardinal rule for the disposal of explosive materials is to do so in such a way as to protect all personnel from the consequences of an explosion that may occur during handling the material. Explosives are prohibited from disposal in landfills, even in a lab pack. Small quantities of commercial explosives [i.e., those classed as DOT Explosives A (see Appendix D)] can be incinerated after desensitizing them by dilution with a flammable solvent or with a flammable solid such as sawdust. Small containers of such diluted explosives should be fed to the incinerator one at a time. The incinerator operator should be fully aware of the nature of the materials being so handled. In general, however, this option is open only to laboratories that have access to their own incinerator facility and also have experts on the handling of explosive materials. Most potentially explosive materials encountered in laboratory work should be transported on public roads only with specialized handling equipment.

One option for disposing of a laboratory material that is potentially explosive is to arrange to have it detonated under carefully controlled conditions. Some laboratories may have personnel who are trained in, and have equipment for, the handling of explosives, and they can remove the material and detonate it on the site where no damage will result. Alternatively, some contract waste-disposal firms have the capability for removing and disposing of explosive laboratory materials. A third possibility is to make arrangements with a local bomb squad or fire department to pick up the material, remove it, and detonate it under safe conditions. In all these situations the chemist should provide the disposer with whatever information is available in the literature on the hazards of the chemical because the disposer may be ignorant of the properties and explosivity of it.

Small quantities of some explosive laboratory materials can be destroyed by procedures given in Chapter 6. However, there are members of each of these classes, as indicated in the following paragraphs, for which laboratory destruction procedures are not applicable. These exceptions should be destroyed by detonation by people trained in handling explosives.

Polynitroaromatic compounds are not easily converted into nonexplosive materials by simple laboratory procedures. Chemical reduction methods, such as reduction with sodium sulfide, tend to be incomplete, and the resulting aminonitro compounds usually still have some explosive properties. A number of the common polynitroaromatic compounds are classified as DOT Explosives A (see Appendix D) and can be disposed of by a contract waste-disposal service that specializes in highly reactive materials.

Picric acid (2,4,6-trinitrophenol) is common in laboratories, where it is often used as a reagent. It is normally sold containing 10-15% water, and in this state it is relatively safe to handle. However, dry picric acid is reported to explode on initiation by friction, shock, or sudden heating.[3] Moreover, picric acid forms salts on contact with metals, and heavy-metal picrates are highly sensitive to detonation by friction, shock, or heat. It is possible that some of the older reported explosions of picric acid may have been initiated by detonation of a minute quantity of a metal picrate in the threads of a metal-capped container. Although picric acid is now sold in plastic-capped containers, it is possible for material in such containers to dry out after repeated opening over a long period of time and thus become hazardous. If picric acid in a plastic-capped container appears to have dried out, the bottle can be immersed upside down in water for a few hours to allow water to wet the threads. The bottle can then be uncapped and filled with water. The water-filled bottle should be allowed to stand a few days to ensure complete wetting of the contents and can then be disposed of by a commercial waste-disposal service or by the local bomb squad or fire department. A metal-capped container of picric acid should be handled only by a trained expert, such as a member of a bomb squad.

Organic peroxides and hydroperoxides, including peroxide-containing solvents, can be treated as described in Chapter 6, Section II.P. However, any peroxidizable solvent from which a solid has crystallized (aged diisopropyl ether is notorious for this) is potentially very hazardous and should be dealt with by people trained in handling explosives.

Moist *diazonium salts* are not explosive and can be converted into disposable material by the coupling procedure of Chapter 6, Section II.Q. However, some diazonium salts are explosive when dry and should be carefully moistened with water before any manipulation is attempted.

Perchloric acid can be disposed of as described in Chapter 6, Section III.C.1. However, ammonium perchlorate and heavy-metal perchlorates are inherently explosive and should be disposed of as explosives. Alkyl perchlorates are treacherously explosive, and their formation should be avoided. If one is formed inadvertently, it should never be isolated, and the mixture in which it exists should be disposed of as an explosive.

Sodium azide can be destroyed by the procedure of Chapter 6, Section III.C.6. However, heavy-metal azides are too treacherous to be handled by this procedure and should be treated as explosives.

Sodium amide and potassium amide can be destroyed by the procedure of Chapter 6, Section III.C.7. However, "many metal derivatives of nitrogenous systems linking nitrogen to a metal (usually but not exclusively a heavy metal) show explosive instability."[4] Compounds with such linkages that are not known to be nonexplosive should be treated as suspect and disposed of as potential explosives.

REFERENCES

1. L. Bretherick, *Handbook of Reactive Chemical Hazards*, 2nd ed., Butterworths, London, 1979, pp. 60–64.
2. NRC Committee on Hazardous Substances in the Laboratory, *Prudent Practices for Handling Hazardous Chemicals in Laboratories*, National Academy Press, Washington, D.C., 1981, pp. 61–72.
3. F. Ullman, ed., *Enzyklopädie der technischen Chemie*, Vol. 4, Urban und Schwarzenberg, Berlin-Vienna, 1929, p. 769.
4. Reference 1, p. 102.

8 Transportation of Hazardous Chemicals

I. INTRODUCTION

Most laboratories are not able to treat or dispose of their waste on the laboratory site. The hazardous wastes must be transported to a legally acceptable landfill, incinerator, or treatment site, often hundreds of miles away. Packaging must be adequate to withstand the physical stress and temperature extremes encountered in transportation.

Generators of waste must comply with a variety of federal regulations, particularly those of the U.S. Environmental Protection Agency (EPA) and the U.S. Department of Transportation (DOT), applicable state regulations, and rules of transporters and disposal facilities. It is important that generators use their knowledge of the properties of the waste to assure safe transport and disposal.

The regulations that cover transportation of hazardous chemicals are extensive and complex. A laboratory that chooses to pack and arrange for transportation of its waste must become thoroughly familiar with these regulations and develop the expertise necessary for compliance with them. Assistance in assuring compliance is available from regulatory agencies, transportation firms, waste-disposal firms, and consultants. EPA has produced a manual[1] to assist waste generators in understanding the interplay among the regulations of the agencies involved. Approved materials, containers, and appropriate labels for shipping hazardous materials from laboratories are commercially available. Laboratories that do not choose to become involved in packing and shipping their waste have the alternative of contracting for the services of a commercial waste-disposal firm to perform the packing and arrange for transportation and disposal of the waste.

It is important to note that movement of hazardous waste in a motorized vehicle from one part of a laboratory site to another may constitute "transportation" under the EPA regulations. If the transporting vehicle moves *along* a street to which the public has free access (as, for example, in many academic campuses), it is considered to be transporting the waste and must do so with required permits and in accordance with DOT regulations. On the other hand, the direct crossing of a public street to go from one building or gate to another is not considered "transportation."

The transport regulations and procedures with which laboratories should be familiar are outlined in the following sections. Additional information on these regulations can be found in Appendix D.

II. THE REGULATORY ENVIRONMENT

The waste generator should have general knowledge of the regulations that govern the transportation of hazardous waste. Since this report is concerned with the disposal of laboratory chemicals, there is no discussion of regulations on transport by tank car, barge, pipeline, or aircraft.

Transportation of chemicals from laboratories is subject to two general areas of federal regulation.

A. U.S. ENVIRONMENTAL PROTECTION AGENCY RESOURCE CONSERVATION AND RECOVERY ACT REGULATIONS

EPA has adopted certain DOT regulations on the transportation of hazardous materials, including rules on shipping descriptions, labeling, marking, placarding, packaging, and the reporting of discharges (see Appendix A, 40 CFR 262 and 263). Moreover, EPA has promulgated regulations with specific requirements for the registration of waste generators, transporters, and disposal facilities. These regulations require the use of a hazardous waste manifest (see Appendix A, 40 CFR 262 and 263) as the principal control for tracking and documenting the movement and ultimate disposal of hazardous waste from the generator to the disposal facility.

EPA also holds generators responsible for assistance in the cleanup of a spill or contamination from a disposal or storage facility. This retained responsibility supercedes any contractual agreement that appears to transfer responsibility to a transporter or disposal firm.

B. DEPARTMENT OF TRANSPORTATION REGULATIONS

DOT has revised its hazardous *materials* transportation regulations to include hazardous *waste* and to regulate intrastate, as well as interstate, transportation. DOT inspectors monitor compliance by transporters. Vehicle drivers are required to know the DOT regulations and to have specified papers, including the manifest, readily available in case of an emergency.

Many states have adopted disposal regulations equivalent to or more stringent than the EPA regulations. A generator who ships waste to another state for disposal may therefore have to comply with the regulations of the states from which and to which the waste is shipped, in addition to those of DOT.

EPA and DOT have jointly proposed the establishment of a uniform manifest (47 FR 9336–9347, 4 March 1982) to eliminate the confusion and compliance problems that have arisen from differing state manifest requirements.

III. RECOMMENDED TRANSPORTATION PROCEDURES

The key elements of safe and legal transportation of waste include the following, which should be included in the laboratory waste management plan (see Chapter 1).

A. HAZARDOUS-WASTE COORDINATOR

It is essential to designate a hazardous-waste coordinator (see Chapter 1, Section IV), who should have primary responsibility for implementing and overseeing all activities related to transportation of hazardous waste from the laboratory.

B. IDENTIFICATION OF REGULATORY REQUIREMENTS

The coordinator must identify and stay abreast of federal, state, and local regulations that control transportation of hazardous waste.

C. SELECTION OF TRANSPORT AND DISPOSAL CONTRACTORS

Waste generators can be held liable for improper transport and disposal of their waste by others and should therefore investigate thoroughly the competence, reliability, and financial stability of transporters and disposal firms with whom they contract for disposal of waste. The hazardous-waste coordinator for the laboratory should not only become assured on these points before entering into a contract with a waste-disposal service but should also check on them periodically during the term of the contract. The coordinator may want to visit the disposal facility to observe its operation.

D. DESIGNATION AND DESCRIPTION OF WASTE CATEGORIES

The coordinator must establish waste categories for the laboratory, based on judgment of safe practices and on regulations and disposal facility rules (see Chapter 2). EPA and DOT regulations require that waste be segregated into categories that can be transported safely; the shipment of some particularly hazardous materials is prohibited (49 CFR 173.21). For transportation purposes each category must be described by the shipping description, which includes the proper shipping name, hazard class, and DOT identification number.

E. PACKAGING OF WASTE

Hazardous waste must be properly packaged for transportation. All containers must be sturdy and leakproof. Inner containers must be cushioned to prevent damage from impact during handling. Outer containers must be properly labeled and marked.

Small containers of laboratory waste are often packed for transport in a lab pack, i.e., surrounded by an inert filler in an outer steel drum (see Chapter 10, Section V.B). This pack is generally used for landfill disposal. An analogous pack in an outer fiber drum can be used for transportation to a waste treatment facility or for incineration. If such a pack is intended for incineration, the incinerator operator may require that the inner individual containers be of linear, high-density polyethylene. A pack in an outer fiber drum is not a "lab pack" as defined by EPA (see Appendix A, 40 CFR 265.316) and cannot be put into a landfill.

F. ARRANGING FOR TRANSPORTATION

EPA controls transporters of hazardous waste by requiring them to obtain an EPA identification number (see Appendix A, 40 CFR 263.11).

If a laboratory is arranging for shipment of its waste, the coordinator must obtain the EPA number of any transporter it employs. Alternatively, the laboratory can employ a contract waste-disposal service that will make arrangements for transportation of the waste.

G. PREPARATION OF SHIPPING PAPERS AND THE HAZARDOUS-WASTE MANIFEST

DOT and EPA both require that shipping papers accompany each shipment of waste. Each agency requires specific information, but neither currently (1983) requires a specific form. Additional state manifests may also be required for intrastate shipment or by both the originating and receiving states in case of interstate shipment. The proposed EPA/DOT uniform manifest would meet the information requirements of both agencies and would eliminate the manifest forms now required by individual states.

H. STORAGE PRIOR TO TRANSPORTATION

Good practices and regulatory requirements for the storage of waste prior to transportation are discussed in Chapter 3.

I. LOADING FOR SHIPMENT

It is prudent for the hazardous-waste coordinator to check the transporter's vehicle for any obvious contamination and for visually apparent defects. All drums or other containers should be reinspected prior to loading to assure that they are not damaged. A final check should be made for appropriate labeling of waste containers. The coordinator should make certain that placards applicable to the hazard priority of the waste are provided.

Drums that contain liquids should be inspected to assure the absence of seepage at seams or bungs. The contents of unacceptable drums should be transferred to acceptable drums, or the drum should be placed inside a DOT-approved recovery drum and packed with absorbent. Care should be taken during transfer operations to avoid inhalation of fumes or skin contact.

J. RECORDKEEPING AND REPORTING

EPA regulations require that the generator keep copies of manifests (40 CFR 262 Subpart D). Copies of any state manifests that may be required should also be retained.

K. PLANNING FOR EMERGENCY RESPONSE

The hazardous-waste coordinator should have an emergency response plan that anticipates possible emergency situations, such as a leaking drum in storage or in transport, a flash fire, or a splash of chemical on a person. Appropriate protective clothing, footgear, face protection, respiratory protection, and absorbent material for liquids should be readily available. Personnel should be trained to deal with the most probable incidents. Among the techniques likely to be needed are the plugging of pinhole leaks in large containers, cleanup of spills, and transfer of liquids from one container to another.

Incidents involving chemicals during and after shipment can pose problems if persons knowledgeable in safe responses to the incident are not readily available. It is essential that vehicle drivers and disposal-facility operators be informed on the hazards of the waste and that appropriate emergency response instructions be available.

Equipment and materials for dealing with some emergencies are commercially available. Information on unfamiliar chemicals can often be obtained by calling the CHEMTREC* hot line (1-800-424-9300). This information service, which is open 24 hours a day, can usually put the caller in contact with a manufacturer of the chemical for information.

REFERENCE

1. *EPA/DOT Hazardous Waste Transportation Interface*, SW-935; EPA Office of Solid Waste and Emergency Response, 1981.

*Chemical Transportation Emergency Center, operated by the Chemical Manufacturers Association.

9 Incineration of Hazardous Chemicals

I. INTRODUCTION

The U.S. Environmental Protection Agency (EPA) regulations define "incinerator" as an enclosed device using controlled flame combustion, *the primary purpose of which is to thermally break down hazardous waste*. This definition differentiates incineration from combustion of waste *primarily for the recovery of its thermal value*. The latter is not considered incineration, and wastes so burned are exempt from incineration regulations (see Appendix A, 40 CFR 261.6).

From an environmental point of view, combustion is probably the method of choice for the destruction of virtually all organic compounds as well as for some wastes that contain inorganic substances. In this chapter we discuss the combustion-destruction of chemicals (1) in an off-site commercial hazardous-waste incinerator, (2) in a laboratory's own on-site hazardous-waste incinerator, and (3) as minor constituents of the fuel feed to a power- or steam-generating plant. The combustion of laboratory wastes in municipal waste incinerators is generally impractical because such incinerators are usually not equipped for or do not have permits for incineration of hazardous wastes. General information on incineration and equipment can be found in References 1–3.

Advantages of incineration, compared with disposal in a secure landfill, include the following:

• With proper emission controls, wastes are converted into innocuous products.
• There is no commitment to long-term containment of hazardous materials.
• Release of contaminants as a consequence of malfunction can be corrected relatively quickly and inexpensively. Malfunction of a secure landfill (e.g., leakage into groundwater) can usually be detected less readily and rapidly and is more costly to remedy.
• Incinerators can handle most reactive wastes that are not allowed in landfills.

On the other hand, incineration has some disadvantages relative to disposal in a secure landfill:

• A costly test burn procedure is required to obtain a permit for a hazardous-waste incinerator.
• Emissions to the atmosphere must be controlled.
• Ash from a hazardous-waste incinerator must usually be finally disposed of in a secure landfill. Such ash is defined by regulation as hazardous unless it is exempted specifically by EPA, based on a showing that it regularly does not meet any of the criteria for a hazardous waste.
• Incineration equipment is relatively expensive to install and to maintain; hence, incineration is more costly to the user than landfill disposal.
• Incinerators operate best on a well-defined feed with known heat of combustion. Incinerator operators are often unwilling to accept the chemically diverse waste from laboratories.

II. CLASSIFICATION AND IDENTIFICATION OF WASTES

Wastes destined for incineration must be classified and segregated by chemical type (see Chapter 2). In addition, incineration characteristics depend on the physical state of the waste, so that classification into organic liquids, aqueous solutions and slurries, oil sludges and bio-sludges, and solids is necessary. Incinerator operators also usually require information on all or most of the following: the chemical identity or chemical class, unusual hazards, approximate heat of combustion, approximate ash content, and chemical compatibility with other wastes.

The bridge between the waste generator and the waste incinerator is the waste-collection system, which should provide for gathering all wastes in a safe manner, segregating and storing them, and arranging for disposal. Conscientious monitoring and recordkeeping are essential for a good collection system.

The minimum size of commercially available (1983) incinerators suitable for hazardous-waste disposal is on the order of 59 kW (200,000 Btu/h) heat duty [e.g., a liquid waste incinerator burning 6.8 kg/h (15 lb/h) of liquid hydrocarbon waste with a heat content of 18,600 kJ/kg (8000 Btu/lb) and 1.82 kg/h (4 lb/h) of natural gas with a heat content of 46,500 kJ/kg (20,000 Btu/lb)].

III. CRITERIA FOR INCINERATOR DESIGN

Most wastes can be destroyed essentially completely (efficiency 99.99%) by high-temperature oxidation. Some oxidation products, such as SO_x, NO_x, HCl, and metal oxides, may be noxious. Emission of SO_x corresponds approximately to the stoichiometric conversion of sulfur in the waste to SO_2, with smaller amounts of SO_3. Nitrogen oxides are formed by oxidation of organically bound nitrogen in the waste, by decomposition of nitrates, and by thermal fixation of atmospheric nitrogen. Halogens in the waste produce a mixture of the hydrogen halide and gaseous halogen; for chlorine-containing material, the predominant product is HCl, which is favored over chlorine at incineration temperatures. Inorganic constituents, depending on their volatility, may either form solid residues or vaporize and recondense to form a submicrometer-sized aerosol. Emissions of the latter may be acceptable, depending on the types and quantities of metal ions they contain, or they may be reduced to acceptable levels by appropriate air-pollution control devices.

Incinerators must be designed to provide the conditions necessary to achieve complete oxidation: a supply of air in excess of the stoichiometric requirement, adequate mixing of air and waste, and sufficiently high temperature to complete oxidation in the time available. These requirements are often designated the 3 Ts of incineration: turbulence, temperature, and time.

A. MIXING REQUIREMENTS

Most small-scale incinerators are used for the disposal of liquid and solid wastes. Incinerators designed for injection of liquids must provide for atomizing the liquids into small enough droplets, generally under 150 μm in diameter, that times for vaporization are short relative to the residence or mixing time in the combustion chamber. Solids in small units are usually supported by either a hearth or a grate. On heating, the solids release volatile products and leave a carbonaceous residue that is subsequently oxidized by reaction with oxygen.

Gross mixing must be achieved between the combustible vapors, which are produced by vaporization of liquids or pyrolysis of solids, and air in order to avoid pockets of gases locally starved of oxygen. In addition, sufficient mixing energy and time must be provided to enable large-scale eddies to break down to a scale small enough to permit mixing to occur on a molecular level. In practice, the gross-scale mixing is achieved by tailoring the injection of waste and air to provide a fairly uniform distribution of both in the combustion chamber, by using high velocities to ensure high turbulence, by use of baffles and cross-jets to promote mixing, and by use of air in excess of the stoichiometric requirement to compensate for imperfect mixing.

B. TEMPERATURE REQUIREMENTS

The high-temperature oxidation of wastes proceeds through a multistep, free-radical process; the factors that control the rate of oxidation are many and not completely understood.[4] The rates increase rapidly with increases in temperature. The oxidation may be inhibited by free-radical scavengers such as halogens.[5,6] Temperatures for the gas-phase reaction range from 1000°C to the adiabatic flame temperature, which may be over 2000°C.

C. TIME REQUIREMENTS

The time required for complete reaction decreases with increasing temperature, from the order of 2 sec at 1000°C to the order of 50 msec at 2000°C. However, the short reaction times attainable at higher temperatures are often precluded by the problems encountered with low-cost refractory materials above 1300°C or with high-alumina refractories above 1500°C.

D. REGULATIONS

Current EPA standards for hazardous-waste incinerators are based on performance.[7,8] The three substantive requirements are a destruction and removal efficiency of 99.99%, a particulate emission below 180 mg per dry standard cubic meter when corrected to 50% excess air, and a requirement on the control of emissions of HCl from the stack of 99% removal or an emission rate of 1.8 kg/h of HCl, whichever is less stringent.

EPA has issued specific requirements that must be met before an incinerator may be operated. These requirements include identification of key fuel elements and principal organic hazardous constituents (POHCs). The destruction and removal efficiency for POHCs must be demonstrated at the 99.99% level by means of trial burns. Removal efficiency refers to the incinerator system as a whole and includes the scrubber as well as the combustion unit.

Certain performance and operating conditions must be monitored, such as fuel rates, temperature, CO level, and oxygen content of the incinerator. The specific requirements vary depending on each specific situation and are an integral part of the procedure for granting permits.

Incinerators used for destruction of hazardous waste must have an EPA permit. The granting of such a permit is contingent on a demonstration by trial burn tests that the incinerator meets EPA performance standards. Trial burn tests may cost up to $100,000 (1982).[9]

The selection of POHCs is based on the ease of destruction and the relative amounts of the materials to be incinerated in a given unit. A permit for an incinerator can be granted on the basis of a trial burn with a contrived waste composed of up to six compounds that are deemed the most difficult to destroy of the compounds for which the incinerator is intended. Once an incinerator has been granted a permit on the basis of such a trial burn, it is deemed qualified to incinerate any compound that is less difficult to destroy than those used in the trial burn contrived waste. EPA has tentatively proposed a hierarchy of relative incinerability based on heats of combustion, but other criteria are also being considered.

E.   RECOMMENDED OPERATING CONDITIONS

The selection of conditions for incinerator operation can be guided by laboratory-scale pyrolysis and oxidation experiments. Several experimental reactors have been developed for determining the ease of destruction or incinerability of chemicals, for possible use in screening chemicals for trial burn selection.[2] The chemicals to be tested are vaporized and premixed with an excess of air and passed through a quartz tube heated to a known temperature in an electric furnace. The destruction efficiency, measured by analysis of the product gases, can be determined as a function of temperature and residence time in the reactor. Tests with such reactors indicate that common organic hydrocarbons, chlorinated organic compounds, nitriles, and amines are at least 99.99% destroyed in 1–2 sec at temperatures of 629–880°C (Table 9.1). These results would apply to an incinerator that depends on the thermal de-

TABLE 9.1 Calculated Temperatures (°C) for Destruction of Selected Compounds with Efficiencies of 99% and 99.99% for Residence Times of 1 and 2 Sec

	Destruction Efficiency			
Compound	99.99%[a] T (°C)	99.99%[b] T (°C)	99%[a] T (°C)	Reference
Acrylonitrile	729	703	—	10
Allyl chloride	691	649	—	10
Benzene	732	717	—	10
Carbon tetrachloride	—	820	—	11
Chlorobenzene	764	744	—	10
Chloroform	—	—	620	11
Dichlorodiphenyltrichloroethane (DDT)	—	—	480	11
2,7-Dichlorodibenzo-*p*-dioxin	—	—	840	11
1-2-Dichloroethane	742	720	—	10
Hexachlorobenzene	—	880	—	11
2,2',4,4',5,5'-Hexachlorobiphenyl	—	—	730	11
Methyl chloride	869	823	—	10
Toluene	727	701	—	10
Triethylamine	594	570	—	10
Vinyl acetate	662	629	—	10
Vinyl chloride	743	722	—	10

[a]Residence time 1 sec.
[b]Residence time 2 sec.

struction of compounds. Differences in the relative ease of destruction would be encountered for incineration in a flame, in which flame-generated oxygen radicals and hydroxyl radicals would increase the destruction rate. Chlorinated compounds are readily degraded thermally because of the weak carbon-chlorine bond but may be difficult to burn because the chlorine radicals inhibit flame reactions.[4-6] In an incinerator, temperatures somewhat higher than the 600–900°C indicated by the small-scale tests would be required to compensate for the time needed to vaporize and mix the chemicals. EPA is investigating different measures of incinerability, and it can be anticipated that data on the destruction efficiency of chemicals subjected to different thermal and oxidation histories will be forthcoming.[12]

However, it is difficult to apply the results of small-scale experiments to full-scale units, and the performance of incinerators is now determined by direct measurement.

IV. PROCESS DESCRIPTIONS

The most common types of incinerators are described in Appendix L by means of process-flow diagrams of representative industrial-size units, which include quench and/or flue gas treatment features. Materials of construction considerations that are important in incinerator systems are also discussed in Appendix L.

The effectiveness of destruction of hazardous wastes in pilot and commercial-scale units of some of the designs described in Appendix L are summarized in Table 9.2. As would be expected, the residence time required for the desired destruction efficiency decreases as the temperature is increased, with effectively complete destruction being attained for hexachlorocyclopentadiene in as little as 0.17 sec at 1378°C.

The designs described in Appendix L are based largely on experience with industrial-scale incinerators. Those designs best suited for adaptation to small scale are, for liquid fuels, blending with a boiler fuel or liquid injection and, for solids, simple hearth or stoker units with one or two chambers. Little information has been published on the performance of small-scale incinerators. A study on the performance of five incinerators carried out by the National Institutes of Health has been reported,[13] and a survey of the availability of small units has been conducted in a study funded by EPA.[14]

V. ECONOMICS

Industrial-scale incinerators are normally purchased on a custom basis rather than off the shelf. Accordingly, it is difficult to obtain even an approximate cost estimate ($\pm 50\%$) for a proposed incinerator; however, some cost data have been compiled for rotary kiln and liquid-waste incinerators.[1] In this analysis the rotary kiln system includes a waste-handling building, rotary kiln combustion chamber, afterburner, water-quench chamber, venturi scrubber, demister, fan, stack, scrubber liquid neutralizer, and associated equipment. The liquid-incineration system includes a waste building, waste storage tank, automated feed system, fuel-oil storage tank, liquid combustion chamber, venturi scrubber, and scrubber water supply and pH control system. Based on their figures and a 10% annual inflation rate, the estimated capital cost for a 293-kW (10^6 Btu/h) rotary kiln would be $1.0 million in 1982 dollars; a liquid waste incinerator of the same capacity would have a capital cost of $200,000. The annual operating costs for these are estimated as $500,000 and $200,000, respectively.

Small-scale incinerators for hazardous waste with capacities down to

TABLE 9.2   Selected Trial Burns in Incinerators[a]

Waste	Incinerator	Temper- ature, °C	Residence Time, sec
Atrazine	Multiple chamber[b]	1070	2.7
		970	2.2
		600	5.6
		970	2.7
		650	6.8
20% DDT in distillate oil; 1.7% PCB in waste oil	Liquid injection; thermal oxidizer[d]	871	NR[c]
		871	3.4
		871	2.9
		982	3.0
		982	4.0
Hexachlorocyclo- pentadiene	Liquid injection[e]	1348–1378	0.17–0.18
Methyl methacrylate (MMA) waste 34% MMA, 13% phenols	Fluidized bed[f]	Bed 774–788	12
		Freeboard 824–843	
Nitrochlorobenzene	Liquid injection[g]	1307	2.3
		1332	2.3
PCB waste	Rotary kiln[h]	Primary 870 Secondary 980–1090	2–3

[a]Abstracted from J. Corini, C. Day, and E. Temrowski, *Trial Burn Data*, draft report, Office of Solid Waste, U.S. Environmental Protection Agency, Washington, D.C., 2 September 1980; see also *Destroying Chemical Wastes in Commercial Scale Incinerators*, final report prepared by TRW under contract 68-01-2966 for the U.S. Environmental Protection Agency, Office of Solid Waste, Washington, D.C., November 1977.
[b]Midwest Research Institute.

about 10 kg/h are commercially available. About 20 manufacturers currently market incinerators with capacities in the range 10–100 kg/h. These units vary widely in design, cost, and performance. The costs for a liquid-waste incinerator or a double-chamber hearth-type incinerator similar in design to those used for the disposal of pathological waste are in the range $50,000–$100,000 (1983). These costs are lower than those that would be obtained by extrapolating the costs cited in the preceding section because small units with emissions of particulates and HCl lower than the limits imposed by EPA do not require pollution-control equipment. Accordingly, these small units are less costly to install and simpler to operate than larger ones. The operating cost for a small incinerator will depend to a large extent on the utilization factor and on the availability at a laboratory site of a skilled operator who could run the

Vaste Feed ate, kg/h	Auxiliary Fuel Rate, kg/h	Excess Air, %	Waste Destruction Efficiency
.63	8.3	43	>99.99
.5	8.5	125	>99.99
.4	2.7	146	>99.99
.3	6.9	71	>99.99
.7	2.5	111	>99.99
)0	0	169	>99.99
)0	0	157	>99.99
!0	0	163	>99.99
30	0	123	>99.99
)0	0	123	>99.99
)-90	120–170	—	>99.99
1800	~3600–4600	11.6–13.4	>99.99
404	~90,000	NR^c	>99.99
350	~80,000	NR^c	>99.99
.5	1780–2250	75–140	99.99

Jot reported.
General Electric Company.
Marquardt Company.
ystems Technology, Inc.
Collins, Inc.
M Company.

incinerator on a part-time basis. A small-volume generator producing 1000 kg/month of waste would need to operate a unit in the above size range for only 10–100 h/month. An additional cost for an incinerator burning a waste of low heating value is that of the auxiliary fuel. For a unit with a capacity of 100 kg/h the fuel requirements could be in the range of 600 kW ($\sim 2 \times 10^6$ Btu/h).

The cost of a trial burn is a major disincentive for the use of small incinerators. There is merit in establishing criteria for small incinerators that are based on design rather than on demonstrated performance. The incinerator standards first proposed under RCRA provided such criteria[15] for halogenated wastes, combustion temperatures over 1200°C, residence times greater than 2 sec, and excess oxygen greater than 3%. These conditions are conservative considering the temperatures and times needed to destroy halogenated compounds given in Tables 9.1 and 9.2. The overall cost of small incinerators would be lowered, and the potential for their use increased, if the requirement for a trial burn were replaced by permit requirements based on conservative design parameters and operating conditions. If the trial burn procedure were still considered necessary, it should suffice to demonstrate performance on the prototype of a standard design and to qualify all units of that design on the basis of the one test. The use of design and operating criteria appears to have been a satisfactory approach to the granting of permits for pathological waste incinerators.

VI. OFF-SITE INCINERATION

Laboratories that do not have their own incinerator may find it cost-effective to arrange for off-site incineration by using a contract waste-disposal service (see Chapters 1 and 8). Some of these services pack laboratory waste and arrange for its transportation and for incineration if appropriate.

The alternative for off-site incineration is to contract with a commercial incinerator operator who is willing to accept the laboratory's waste. To follow this route the laboratory must identify and segregate its waste as required by regulations (see Chapter 2, Appendixes A and D) and arrange for transportation to the incinerator (see Chapter 8). If the contract with the incinerator operator does not specify types of waste to be accepted and costs of handling, it is prudent to get the operator's agreement on accepting a proposed shipment and a cost estimate before shipping to the facility. Many commercial incinerator operators do not accept unsegregated waste or waste packed like a lab pack in a fiber drum.

VII. USE OF EXISTING BOILERS

Some wastes, such as liquid hydrocarbons and their oxygenated derivatives, can be burned for heat recovery as secondary fuels in steam boilers; this practice is not considered "incineration" under EPA regulations. Such wastes are usually mixed directly with the primary fuel, preferably in low proportion, so as not to make a significant change in the boiler operation. Local regulations and/or the terms of a boiler's operating permit may restrict the burning of waste chemicals in a particular boiler.

Present RCRA regulations exempt steam boilers that burn qualified wastes for heat recovery from the requirement of 99.99% destruction and removal efficiency. However, there are regulatory constraints on the types of waste that qualify for burning for recovery of heat value. EPA guidelines[16] state that burning organic waste that has little or no heat value in industrial boilers under the guise of energy recovery is not within the exemption for recycling. These guidelines prescribe that a waste that qualifies for legitimate heat recovery in a boiler should have a higher heating value (heat of combustion) than that of wood or low-grade subbituminous coal: about 18600 kJ/kg (8000 Btu/lb). This limitation excludes polyhalogenated compounds; many polynitro aromatics; and many compounds that contain significant amounts of nitrogen, phosphorus, or sulfur. A list of some hazardous compounds that have heats of combustion less than this limit is given in Table 9.3.

VIII. ON-SITE INCINERATION

The successful destruction of hazardous wastes by incineration has been demonstrated both in large-scale facilities, many of which have the capability of accepting a variety of wastes, and in small units designed to handle wastes of known and fairly constant composition. There are problems, both technical and institutional, that need to be addressed when considering acquiring and operating an incinerator for laboratory waste that comprises relatively small amounts of a wide variety of chemicals. Some of these problems are discussed in this section.

An important question in the handling of packaged wastes is whether it is preferable to burn them in their packages and accept the need for destroying the package or to transfer them to storage containers with similar compounds to take advantage of handling larger lots. Combustion of solvents is best accomplished by the continuous flow of liquid of constant composition from a holding tank. Safe procedures must be established to protect operators from exposure to hazardous chemicals during transfer of waste into holding containers and blending with fuel oil. Chemically incompatible materials (see Appendix E) must not be mixed in such operations.

Incinerators must usually be operated within a relatively narrow temperature range, with the lower end established by the temperature needed to ensure adequate destruction of the waste (e.g., 1200°C for some halogenated compounds) and the upper bound by materials of construction (as low as 1300°C for some refractory surfaces). A burner fired with auxiliary fuel can provide the energy both for preheating an incinerator to the desired temperature levels and for maintaining it at temperature when handling wastes of low heat content. However, the potential exists for rapid temperature excursions, and possibly explosive surges of gas

TABLE 9.3   Low-Energy Hazardous Constituents Listed in 40 CFR 261, Appendix VIII[16]

Hazardous Constituent	Higher Heating Value (Btu/lb)
Tribromomethane	234
Tetrachloromethane	432
Hexachloroethane	827
Dibromomethane	899
Pentachloroethane	953
Hexachloropropene	1,259
Chloroform	1,349
Cyanogen bromide	1,457
Trichloromethanethiol	1,475
Tetrachloroethene (tetrachloroethylene)	2,141
Cyanogen chloride	2,320
Iodomethane	2,410
Tetrachloroethane, n.o.s.	2,500
1,1,1,2-Tetrachloroethane	2,500
1,1,2,2-Tetrachloroethane	2,500
1,2-Dibromoethane	2,572
1,2-Dibromo-3-chloropropane	2,682
Bromomethane	3,058
Dichloromethane	3,058
Trichloroethene (trichloroethylene)	3,130
Hexachlorobenzene	3,220
Bis(chloromethyl) ether	3,544
1,1,1-Trichloroethane	3,580
1,1,2-Trichloroethane	3,580
Pentachlorobenzene	3,688
Pentachlorophenol	3,760
Hexachlorocyclopentadiene	3,778
Hexachlorocyclohexane	3,813
Kepone	3,867
2,3,4,6-Tetrachlorophenol	4,011
Endosulfan	4,191
1,2,4,5-Tetrachlorobenzene	4,695
Bromoacetone	4,785
Dichloroethylene, n.o.s.	4,857
1,1-Dichloroethylene	4,857
Chlordane	4,875
Heptachlor epoxide	4,875
Phenylmercury acetate	4,875
Acetyl chloride	4,983
Trichloropropane, n.o.s.	5,055
1,2,3-Trichloropropane	5,055
Dichloropropanol, n.o.s.	5,109
Dimethyl sulfate	5,145
2,4,5-T	5,163
2,4,5-Trichlorophenol	5,181
2,4,6-Trichlorophenol	5,181

TABLE 9.3 *(continued)*

Hazardous Constituent	Higher Heating Value (Btu/lb)
N-Nitroso-N-methylurea	5,196
1,1-Dichloroethane	5,396
1,2-Dichloroethane	5,396
trans-1,2-Dichloroethane	5,396
Phenyldichloroarsine	5,612
N-Nitrososarcosine	5,738
Azaserine	5,774
2-Fluoroacetamide	5,828
Benzenearsonic acid	6,116
Maleic anhydride	6,116
1,2,4-Trichlorobenzene	6,116
TCDD	6,170
Dichloropropene, n.o.s.	6,188
1,3-Dichloropropene	6,188
Endrin	6,224
Trinitrobenzene	6,224
Chloromethyl methyl ether	6,260
2,4-Dinitrophenol	6,332
Nitrogen mustard N-oxide and hydrochloride salt	6,404
Parathion	6,494
2,4-D	6,512
1,3-Propane sultone	6,602
Methyl methanesulfonate	6,728
Aldrin	6,746
Nitroglycerine	6,818
2,4-Dichlorophenol	6,854
2,6-Dichlorophenol	6,854
Hexachlorophene	6,871
Trypan blue	6,907
Benzotrichloride	7,015
Cycasin	7,105
N-Nitroso-N-ethylurea	7,105
Cyclophosphamide	7,141
Methylparathion	7,145
Uracil mustard	7,145
1,2-Dichloropropane	7,171
Dichloropropane, n.o.s.	7,178
Amitrole	7,213
Dimethoate	7,231
Tetraethyllead	7,267
4,6-Dinitro-o-cresol and salts	7,303
N-Methyl-N-nitro-N-nitrosoguanidine	7,303
Mustard gas	7,303
Dinitrobenzene, n.o.s.	7,465
N-Nitroso-N-methylurethane	7,519
Nitrogen mustard and hydrochloride salt	7,699
Hydrazine	7,987

evolution may occur when burning liquids of high volatility if precautions are not taken to feed the liquids at low rates. Closed metal containers introduced into incinerators may explode because of high pressure buildup in them.

Cold surfaces in an incinerator will quench combustion reactions and lead to emission of unburned or partially burned materials. This potential source of undesirable emissions is greatest in small units, which have a high surface-to-volume ratio. The problem of wall quenching can be minimized by using hot refractory surfaces. Waste should not be fed to an incinerator during periods of startup or shutdown, when the walls are below operating temperature.

There is a large challenge, particularly for small institutions with limited resources, in selecting an incinerator of the scale and design to meet the specialized needs of a given laboratory. Little documentation is available on experience with the use of small-scale on-site incinerators for the disposal of hazardous wastes from laboratories.

REFERENCES

1. T. Bonner et al., Hazardous Incineration Engineering, *Pollution Technology Review*, No. 88, Noyes Data Corporation, Park Ridge, N.J., 1981.
2. U.S. Environmental Protection Agency, *Engineering Handbook for Hazardous Waste Incineration*, Draft, Cincinnati, Ohio, November 1980.
3. M. Sittig, Incineration of Industrial Hazardous Wastes and Sludges, *Pollution Technology Review*, No. 63, Noyes Data Corporation, Park Ridge, N.J., 1979.
4. A. Tsang and W. Shaub, "Chemical processes in the incineration of hazardous materials," presented at 182nd National Meeting of the American Chemical Society, New York, August 1981.
5. D. Bose and S. M. Senkan, "On the mechanism of chlorinated hydrocarbon combustion," *Combustion Science and Technology*, in press.
6. W. E. Wilson, J. T. O'Donovan, and R. M. Fristrom, in *Twelfth International Symposium on Combustion*, The Combustion Institute, Pittsburgh, Pa., 1969, pp. 929–942.
7. 46 FR 7666, 23 January 1981.
8. 47 FR 27516, 24 June 1982.
9. *Chemical and Engineering News*, p. 7, 28 June 1982.
10. K. Lee, N. Morgan, J. L. Hansen, and G. M. Whipple, "Revised model for the projection of the time-temperature requirements for thermal destruction of dilute organic vapors and its usage for predicting compound destructability," presented at the 75th Annual Meeting of the Air Pollution Control Association, New Orleans, La., June 1972.
11. B. Dellinger, University of Dayton Research Institute, personal communication, November 1982.
12. D. S. Duvall, W. A. Rubey, and J. A. Mescher, in *Treatment of Hazardous Wastes*, Proceedings of the Sixth Annual Research Symposium (D. Schultz, ed.), EPA-600/9-80-011, U.S. Environmental Protection Agency, Cincinnati, Ohio.

13. T. K. Wilkinson and H. W. Rogers, Chapter 30 in *Safe Handling of Chemical Carcinogens, Mutagens, and Teratogens, and Highly Toxic Substances*, D. B. Walters, ed., Ann Arbor Science Publishers, Ann Arbor, Mich., 1980.

14. V. S. Engleman and D. L. deLesdernier, "Incineration technology for selected small quantity hazardous waste generators," presented at 8th Annual Research Symposium on Solid and Hazardous Waste Research and Development, Fort Mitchell, Ky., 8–10 March 1982.

15. 43 FR 59008, 18 December 1978.

16. 48 FR 11157–11160, 16 March 1983.

10   Disposal of Hazardous Chemicals in Landfills

I. INTRODUCTION

Waste that must be disposed of off the laboratory site is usually incinerated or put into a landfill; a substantial fraction of laboratory waste now goes to landfills. This situation is not likely to change soon because many inorganic wastes and some organic wastes cannot be incinerated, and commercial incineration facilities are not accessible to many generators of laboratory waste. Laboratory wastes that are known to be nonhazardous are suitable for disposal in a sanitary (e.g., municipal) landfill. However, the preponderance of laboratory wastes is either hazardous and regulated by the U.S. Environmental Protection Agency (EPA) or is properly considered hazardous even though not regulated by EPA. This latter class of hazardous waste includes compounds that do not meet any of the four EPA hazard characteristics and that are not specifically listed by EPA (Appendix A, 40 CFR Subparts C and D). An example would be a newly synthesized compound that cannot be inferred to be nontoxic from its structure and that must therefore be assumed to be toxic. Landfill disposal of either type of hazardous waste must be in a secure landfill approved by EPA or the state.

The advantages of disposal in a secure landfill, compared with incineration in a commercial incineration facility, include the following:

126

- A secure landfill is generally more accessible to laboratories than is a commercial incinerator. Even though hazardous waste may have to be transported a considerable distance to a secure landfill, many commercial incinerator operators will not accept laboratory waste because of its chemical diversity, relatively low volume, and requisite paperwork that is disproportionate to its volume.
- Relatively low cost, which is a reflection of the fact that a secure landfill is less costly to construct and to operate than a hazardous-waste incinerator.
- Requirements for segregation of wastes are less severe. Some types of waste that can be packed together for landfill disposal must be segregated for incineration (e.g., halogenated compounds must be kept separate for incineration).
- Flexibility in the types of waste that can be disposed of at a given time. Incineration of laboratory wastes may depend on what other types of waste the incinerator is handling.

On the other hand, secure landfill disposal has disadvantages relative to incineration.

- A secure landfill must be committed to long-term containment of waste. This commitment requires careful attention to design features that can reduce the probability of groundwater contamination from leaks.
- Repairing a loss of containment is costly, and the cost may well be reflected in costs to users.
- Energy recovery is currently not possible.

The disposal of laboratory waste in a secure landfill is now, and may continue to be for some time, the only practical option for disposal of many laboratory wastes. For reasons of economy, wastes destined for landfill should be concentrated to the extent feasible. Some ionic solutions can be converted into low-volume solids by precipitation (see Chapter 6), and others can be concentrated by evaporation of water.

II. SANITARY LANDFILL—NONHAZARDOUS LABORATORY WASTES

Some solid chemical wastes that are produced in laboratories are not hazardous by any definition. While it may be expedient to treat these materials in the same manner as hazardous wastes, it is important to reduce the volume of hazardous waste as much as possible. These nonhazardous wastes are suitable for disposal in a sanitary landfill and can be put into one if local regulations permit. If sanitary landfill disposal

TABLE 10.1   Typical Nonhazardous Laboratory
Wastes

A. *Organic Chemicals*
   Sugars and sugar alcohols
   Starch
   Naturally occurring α-amino acids and salts
   Citric acid and its Na, K, Mg, Ca, NH_4 salts
   Lactic acid and its Na, K, Mg, Ca, NH_4 salts
B. *Inorganic Chemicals*
   Sulfates: Na, K, Mg, Ca, Sr, Ba, NH_4
   Phosphates: Na, K, Mg, Ca, Sr, NH_4
   Carbonates: Na, K, Mg, Ca, Sr, Ba, NH_4
   Oxides: B, Mg, Ca, Sr, Al, Si, Ti, Mn, Fe, Co, Cu, Zn
   Chlorides: Na, K, Mg
   Fluorides: Ca
   Borates: Na, K, Mg, Ca
C. *Laboratory Materials Not Contaminated with Hazardous*
      Chemicals
   Chromatographic adsorbent
   Glassware
   Filter paper
   Filter aids
   Rubber and plastic protective clothing

is available to a laboratory the laboratory waste-disposal plan can provide for separate containers, marked "Nonhazardous Waste," to assure
waste handlers of their safety and to guide these materials to sanitary
landfill disposal.

Table 10.1 lists examples of nonhazardous wastes that are suitable for
disposal in a sanitary landfill. These materials are not hazardous according to EPA criteria and are of sufficiently low toxicity for safe
disposal in a sanitary landfill. All the materials on this list have $LD_{50} >$
500 mg/kg. Other criteria for toxicity may be used, depending on state
or local regulations. The laboratory waste-disposal plan should list those
materials pertinent to its operation that are not hazardous.

III.   IN-HOUSE DETOXIFICATION OF HAZARDOUS WASTES FOR LANDFILL

Some hazardous laboratory wastes can be converted into nonhazardous
wastes by treatment in the laboratory, resulting in residues suitable for
disposal in a sanitary landfill. This course of action reduces the burden
on a secure landfill as well as the cost of using one and should be pursued

whenever the type and quantity of wastes make it feasible. Procedures for laboratory destruction-detoxification of many types of common laboratory chemicals are given in Chapter 6.

IV. COMMERCIAL SERVICES

There are commercial firms that provide contract service for disposal of laboratory wastes. Some of these firms have chemically knowledgeable employees who can prepare lab packs (see this chapter, Section V.B) of laboratory wastes (and some supply their own packing materials) and arrange for transportation of the lab packs and disposal of the wastes. Costs for such service (1982) are in the range of $200 per lab pack. A laboratory wishing to use such a service must have a waste collection and segregation system that conforms to the contractor's requirements. It should also be borne in mind that the generator of hazardous waste is responsible for safe and legal disposal of the waste, even when the disposal is done by others. Accordingly, a laboratory using such a service should investigate and remain aware of the eventual disposal procedure (see Chapter 8, Section III.C).

It is sometimes possible to ship laboratory wastes to a commercial treatment facility for eventual landfilling. However, such facilities are designed to handle large quantities of individual wastes from manufacturing plants by treating them chemically or physically to convert them into wastes that can be landfilled. Accordingly, they are often unwilling to deal with laboratory wastes because of the uneconomically small quantity and because of their need to know the composition of wastes to be treated. Commercial treatment facilities generally require an analysis of the wastes they receive. If a single barrel contains a large number of discrete chemical entities, analysis is completely infeasible, and the result is that the wastes are not accepted.

V. DISPOSAL OF HAZARDOUS WASTES IN A SECURE LANDFILL

A. PRESENT REGULATORY STATUS

Hazardous wastes of Small Quantity Generators (see Appendix A, 40 CFR 261.5) are exempt from Resource Conservation and Recovery Act regulations except for the requirements for proper treatment, storage, transportation, and disposal; allowed methods of disposal include disposal in a state-licensed municipal (sanitary) landfill [see Appendix A, 40 CFR 261.5(g)(3)(iv)]. Some states do not accept this option for disposal of small quantities of wastes and have either eliminated the Small Quantity Exemption or at least do not permit any quantity of hazardous waste to be disposed of in municipal landfills (see Appendix B). Even in states where Small Quantity Generators can legally send hazardous wastes to municipal landfills, this is not a desirable practice because the handling, disposal, and fate of the wastes is under little or no control.

Accordingly, the need for laboratory access to secure landfills is imperative. EPA recognized this fact in an amendment (the lab pack amendment; see Appendix A, 40 CFR 265.316) to EPA regulations on landfill disposal. These regulations ban land disposal of all ignitable, reactive, or liquid waste, even in secure landfills. The lab pack amendment exempts from the ban "small containers of hazardous waste in overpacked drums (lab packs)." The lab pack process is specified in this amendment and is discussed below in Section V.B.

Other regulatory requirements that affect land disposal of laboratory wastes are those placed on owners and operators of treatment and disposal facilities. These facilities are required to maintain records and to file biennial reports (see Appendix A, 48 FR 3977–3982, 28 January 1983) detailing all wastes handled by EPA Hazardous Waste Number [Appendix A, 40 CFR 261.30(c)] and by quantity. A single lab pack of laboratory wastes may be accompanied by a manifest listing many discrete items. The result is a paperwork burden on the facility operator that may be accepted reluctantly; some facility operators simply choose not to handle laboratory waste at all.

Because of the characteristics of, and regulatory requirements placed on, laboratory wastes—small quantities, large variety, need for characterization of each waste, and extensive recordkeeping—there is a general deficit of secure landfill disposal facilities for laboratory wastes in many parts of the country. Some wastes are being transported more than 1000 miles for landfill disposal.

More recently, EPA has issued regulations that set forth criteria for the design, construction, and operation of secure landfills. These regulations may result in increased costs to users of secure landfills, although their impact on continuing operation of existing landfills and on the establishment of new ones is yet to be seen. It is likely that a laboratory that wishes to establish its own secure landfill for even the relatively small quantities of its wastes would find the costs of constructing and operating it, and the requisite recordkeeping, economically impractical.

B.   THE LAB PACK METHOD (See Appendix A, 40 CFR 265.316)

The most common method of secure landfill disposal of laboratory wastes is the use of a 55-gallon open-head steel drum that is filled with small containers of chemicals packed in and separated by an absorbing medium. The requirements for a lab pack are summarized below.

All lab packs must be in U.S. Department of Transportation (DOT) specification outside drums, and all outside drums must be metal. The drums must be of the open-head variety to allow the proper placement of the inside containers and absorbent. Inside containers are not specified but must not react dangerously with, be decomposed by, or be ignited by the waste held therein. Inside containers must also be nonleaking and tightly and securely sealed.

The inside containers must be overpacked and surrounded, at a minimum, by a sufficient quantity of absorbent material to completely absorb all the liquid contents of the inside containers. In addition, the outside container must be full after packing with the inside containers and absorbent material to prevent breakage of inside containers. The absorbent material used must not be capable of reacting dangerously with, being decomposed by, or being ignited by the contents of the inside containers. It appears that vermiculite and fuller's earth are commonly used because of their price, availability, and the fact that they will not react dangerously with most wastes. Commercial absorbents are available that have greater absorptive capacity than vermiculite or fuller's earth.

EPA prohibits the placement of incompatible wastes (see Appendix E) in the same outside container. The purpose of this restriction is to prevent any potentially dangerous reaction between wastes packaged in the same lab pack. Reactive wastes that can explode or release toxic gases, vapors, or fumes when they are at standard pressure and temperature; when they are mixed with or exposed to water; when they are subjected to a strong initiating force; or when they are heated under confinement or else are DOT-forbidden, Class A, or Class B explosives (see Appendix D, Table D.2) are banned from landfill disposal. Cyanide- and sulfide-bearing wastes are permitted in lab packs.

The stipulation that incompatible wastes not be packed together is usually met by using DOT classes and shipping requirements (see Appendix D) to designate the contents of containers (e.g., one drum for flammable, one for corrosive). Further segregation requirements may be placed on the generator by the shipper or land-disposal facility. Persons responsible for sorting and packing of chemicals for land disposal should have a decision guide available to help determine the appropriate category for each waste; a simplified example of such a guide is shown in Figure 10.1. Specific criteria such as pH and flash point may need to be adjusted to comply with state or local regulations or disposal-facility requirements.

The inclusion in Figure 10.1 of a decision criterion based on toxicity reflects the fact that many laboratory chemicals that are not listed by EPA (see Appendix A, 40 CFR Subparts C and D) are known or suspected to be toxic; these materials should be handled as hazardous waste even though they are not EPA-regulated hazardous waste. If the toxicity of a chemical is not known or cannot be reasonably inferred by analogy to similar compounds, it should be assumed to be toxic. The specific limits for toxicity given in Figure 10.1 may vary from state to state or laboratory to laboratory, but provision should be made to control the disposal of presumably toxic but unregulated chemicals. The information on the characteristics of waste that is necessary for the use of such a decision guide must be provided by the laboratory worker who generated the waste.

VI. OTHER METHODS OF PREPARING WASTES FOR LANDFILL DISPOSAL

Although the lab pack is the most generally useful method for handling chemicals for landfill disposal, other methods can be used and in some situations may be preferred.

A. SOLIDIFICATION OF LIQUID WASTES BY ABSORPTION

Liquid wastes can be solidified by combining them with sufficient inert absorbent so that no free liquid remains, and a steel drum filled with such a mixture can be disposed of in a secure landfill. The procedure can be useful for handling wastes that cannot be disposed of by incineration or other combustion and that are of too large volume for the lab pack method. A small-scale check should be made to determine that the absorption process is not strongly exothermic. The absorbents should be the same types as those used in lab packs (this chapter, Section V.B).

This method cannot be used for any of the types of reactive materials that are prohibited from landfill disposal (this chapter, Section V.B) or for materials that have the EPA hazard characteristic of corrosivity. Only compatible liquids (see Appendix E) should be so treated for packaging in a single drum.

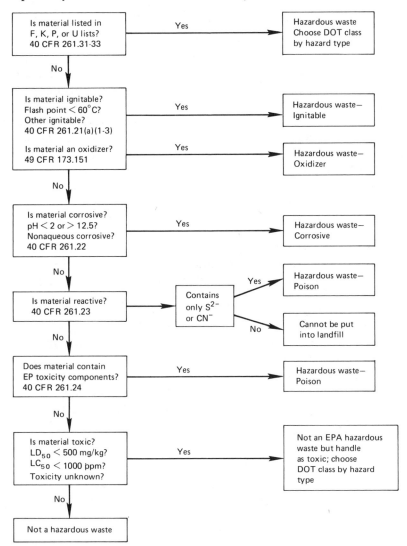

FIGURE 10.1   Decision guide for sorting laboratory waste for landfill disposal. (See Appendix A for references to CFR.)

References for Figure 10.1: *The Condensed Chemical Dictionary*, G. G. Hawley, ed., 10th ed., Van Nostrand Reinhold, New York, 1981; *The Merck Index*, Merck and Co., Rahway, N.J., 9th ed., 1976; *NIOSH Registry of Toxic Effects of Chemical Substances*, U.S. Government Printing Office, Washington, D.C., 1980; *Dangerous Properties of Industrial Materials*, 4th ed., Van Nostrand Reinhold, 1979; *Aldrich Catalog*, Aldrich Chemical Co., 1982–1983; *Sigma Catalog*, Sigma Chemical Co., 1983; *Pesticide Manual*, British Crop Assoc., 4th ed.; L. Bretherick, *Handbook of Reactive Chemical Hazards*, Butterworths, London-Boston, 1979.

B. ISOLATION OF WASTES BY ENCAPSULATION

Some wastes are best treated by isolating them from the environment, and a number of procedures have been developed for doing so. Some of these procedures are proprietary and if carried out by a commercial treatment facility can be quite expensive for the small generator. However, some encapsulation techniques can be performed in-house with small quantities. The resulting waste exhibits no hazard characteristic or leachability and can therefore be disposed of in a sanitary landfill.

One encapsulation technique is to incorporate the waste into a matrix of an inert, nonleachable solid. Some wastes can be mixed with cement and water to form a concretelike solid that can be put into a landfill. Polymers and gels can be purchased that, when mixed with wastes, including liquids, set up to form a nonleachable solid that can be landfilled.

Another approach is to encapsulate containers of wastes inside a concrete block.[1] These blocks should be relatively small for ease of handling and to ensure structural integrity—they can be formed inside empty metal 5- or 15-gallon containers, which give satisfactory sizes and can also solve the problem of disposing of the containers. Care should be taken that the concrete mixture forms a strong concrete block; reinforcement with metal wire or rods will increase the strength. Waste containers so encapsulated should not touch one another or come within 5 cm of the outside of the block. Only full containers of wastes should be used in order to avoid voids in the block should the individual containers break. Blocks of this type, properly prepared, can be put into many landfills.

REFERENCE

1. J. Meister, *Operation of the Southern Illinois University (Carbondale) Waste Program,* Conference on Waste Management in Universities and Colleges, Association of Physical Plant Administrators, Suite 250, 11 Dupont Circle, Washington, D.C. 20036.

11 Disposal of Chemically Contaminated Waste from Life-Science Laboratories

I. INTRODUCTION

Life-science laboratories vary widely in the kinds of activities carried out in them and, hence, in kinds of waste. The waste may be solid or liquid and includes such materials as carcasses, tissues, animal bedding and excrement, plant material, and microorganism cultures. Any of these may be contaminated with hazardous chemicals, depending on the laboratory activity. The microorganisms (e.g., bacteria, fungi, viruses) may be hazardous pathogens.

Purely chemical operations are often carried out in life-science laboratories. The proper disposal of waste from such operations is described elsewhere in this report. This chapter deals with biological waste with or without hazardous chemicals present. However, the emphasis is on waste that is chemically contaminated. This is a waste category that, unlike purely chemical waste or purely biological waste, is virtually unaddressed by either scientific literature or regulations.

Procedures for handling and disposal of biological waste, which range from the disinfection or sterilization of contaminated materials to the incineration of animal carcasses, are well documented.[1-3] Specific and generally accepted protocols for the removal and disposal of biological waste and its associated contaminated materials have evolved from years of experience and research in academic, industrial, and government laboratories (see the references at the end of this chapter). One unifying theme in these documents is the emphasis on careful assessment of the type of biological waste to ensure the appropriate selection of a waste-disposal strategy (e.g., decontamination, disinfection, sterilization, incineration).

Reliance on autoclaving, incineration, or burial as a treatment or disposal strategy is prevalent, although in some cases inappropriate. While generally recognized as standard procedures, their efficacy is not universal and their application is not always necessary.* Prudent, effective, and less costly alternative handling techniques exist for innocuous as well as for hazardous biological wastes.† It is the responsibility of investigators to characterize their biological wastes and judiciously apply known techniques, materials, and equipment to minimize the hazard to health and environment and the cost to the laboratory.

II. DEFINING THE WASTE

The first step in managing any waste is to characterize it. For chemically contaminated biological waste (CCBW) this step consists of identifying types and sources of the waste; determining its rate of generation; and describing its physical, chemical, and biological properties.

Typical examples of CCBW from life-science laboratories include dead animals and animal tissues, animal bedding and excrement, cell cultures, and body fluids. The chemical contaminant usually results from administration of chemicals to determine their effects on life systems or application of chemicals for a germicidal or disinfecting purpose. Typically, the mass of the chemical or drug is quite small in comparison with the mass of the biological substance. For example, an animal challenged with a suspected carcinogen might contain only micrograms or milligrams of the challenge material. Similarly, plants used in agricultural research laboratories may be treated with minute amounts of insecticide or fungicide. Because the apparent physical properties of the combined CCBW are usually the physical properties of the biological material alone, it is quite common to manage CCBW in the same manner as

---

*Heat-resistant spores, animal carcasses, or some biological toxins (e.g., aflatoxin) may not be effectively treated through the use of autoclaving.

†Disinfectants that contain active chlorine or iodine are useful for killing bacteria, bacterial spores, viruses, rickettsiae, and fungi on large pieces of equipment subject to surface contamination or in heat-sensitive materials whose subsequent use or effectiveness might be compromised if autoclaved.

purely biological waste is managed. As will be discussed further, such a waste management assumption is not always advisable.

Before a treatment and/or disposal strategy can be selected, consideration must be given to the nature of the chemical contaminant associated with the biological material. The following questions should be answered before selecting the most effective treatment/disposal process:

1. What is the chemical contaminant, and how much was added to the biological material?

2. What is the fate of the contaminant within the biological system? How much chemical is metabolized or otherwise altered, and how much remains unchanged? Are more hazardous reaction products formed?

3. What are the known toxic properties of the chemical contaminants or reaction products?

4. Are the chemicals or reaction products easily volatilized, oxidized, or otherwise affected by the input of some form of energy?

5. Does the waste contain radioactive material? Disposal of such waste is strictly regulated.[4]

Although answers to the above questions will not always be readily available, they should be earnestly sought in order to minimize risks from the selection of an inappropriate treatment or disposal technique.

III. TREATMENT/DISPOSAL TECHNIQUES

It is a basic principle that viable organisms in biological materials must be killed before they leave the laboratory. This is true even if their destination is an incinerator, for they may contaminate the ash or be carried off in the gases if incineration is not rapid and complete; it applies particularly to certain viruses and spore-forming organisms. For animal tissue and excrement containing their normal microbial population, direct incineration is usually acceptable. However, if the material may contain disease-causing organisms, autoclaving or other treatment should precede incineration. With proper selection of temperature, pressure, and time, autoclaving is often highly efficient. Leading references to the voluminous literature on this important subject are listed at the end of this chapter. Treatment is usually achieved by one of the following approaches:

1. Heat/pressure, such as autoclaving, wet heat, dry heat.

2. Chemicals (wet/gaseous), such as chlorine, iodine, phenols, acidified alcohols, ethylene oxide, ozone, formaldehyde, or various quaternary ammonium salts.

Ultimate disposal, following treatment, is usually to the sewer system for certain liquids or to a sanitary landfill for solids. Treatment products can include autoclaved tissue culture fluids, ash and residue from a pathological waste incinerator, and pipettes that have been soaked in a chemical disinfectant such as an iodine solution. Local regulations, in addition to those of the federal and state governments, may control wastes discharged to both sewage systems and sanitary landfills, and consequently regulations at all these levels should be consulted before selecting a final disposal strategy for such products.

If the biological waste contains a chemical contaminant, the selection of a treatment/disposal strategy becomes more complicated than when considering only biological waste. For example, autoclaving a cell culture contaminated with a highly volatile carcinogen could lead to a dangerous release of the carcinogen. In incineration, a common treatment for dead animals, the volatilization of the chemical contaminant may be beneficial, particularly if the chemical itself is oxidized to harmless end products. However, if the contaminant contains a volatile metal, such as mercury or lead, or is an organic compound that is difficult to destroy by combustion, such as tetrachlorodibenzodioxin, incineration may actually serve to disperse a toxic substance into the air, and some other method of disposal should be chosen. Similarly, for chemical decontamination of pathogenic materials that contain added chemicals, an assessment of the interaction between the chemical species is helpful in anticipating potentially harmful end products (e.g., chlorination yielding chloramines) and in assuring that the amount of disinfecting chemical is adequate to destroy the pathogens.

In some cases considerations such as those raised above will help to identify a treatment method that satisfactorily addresses both chemical and biological concerns. It is quite possible, however, that the chemical contaminant is in such high concentration or is of such a nature that traditional biological treatment methods are not suitable. In such cases it may be necessary to give priority to the chemical nature of the CCBW and treat it as a hazardous waste. Often the CCBW can be treated and rendered innocuous by chemical waste incineration.* Otherwise, disposal in a secure landfill may be the solution.

---

*Multipurpose chemical-waste incinerators are usually operated under more rigorous combustion conditions than pathological incinerators. See Chapter 9 for additional discussion of this subject.

IV. INCINERATION

Incineration, already widely used to dispose of biological wastes with or without chemical contamination, is growing in importance; hence, some detail on its use is justified.

Incineration is a clean, efficient method of disposing of animal carcasses, animal organs, animal excrement and the paper or other bedding on which it is collected, and wood or cardboard containers in which animals have been shipped. The preferred type of incinerator for carcasses and organs is the pathological incinerator, which has a solid hearth rather than a grate in the ignition chamber. A disadvantage of the grate is that small carcasses and organs may drop through it without being destroyed. Moreover, the hearth facilitates release of the water that makes up about 85% of animal tissue and that must be evaporated before destruction of organic matter can take place. Location of primary and secondary burners in the incinerator is critical; flames must impinge directly on the wastes to maximize evaporation and combustion. Appropriate incinerators are described in Chapter 9.

There are good reasons to have the incinerator on the laboratory site if local ordinances permit it and if there is enough incinerable material to make the unit economical. The advantages are that direct control of the operation helps assure proper incineration and that less documentation is required: DOT regulations do not apply to intrasite transportation (see Chapter 8, Section II).

The argument for incineration is stronger for animal material containing hazardous chemicals than for chemical-free animal material, as there are relatively few landfills approved for hazardous chemicals and they are likely to be so far away that shipping costs are high. Wet animal waste, such as carcasses and excrement, should be placed in bags of plastic, such as polyethylene. The bags should be tied or taped shut. If a bag is to be incinerated on the site, it is placed in a sealed cardboard carton or in a heavy laminated paper bag. Paper bags should be folded at the top and stapled shut. For ease of handling the loaded bags should not weigh more than about 20 kg (44 lb). Unless bags containing carcasses are collected and incinerated daily, they should be stored in walk-in refrigerators maintained at about 4°C (40°F).

Glass and metal should be excluded from bags going to a hearth-type incinerator because of the possibility of explosion of closed bottles or cans and increased incinerator maintenance due to fused glass or excessive ash. Appropriate labels should be attached, such as "Animal Carcasses" and "Caution—Cancer Suspect Agent."

If the material is to be incinerated off site, packaging and labeling must be in accord with DOT regulations (see Chapter 8).

A laboratory with access to good incineration facilities should not assume that all its biological waste can or should be incinerated; it is likely that some of the laboratory's biological waste should be treated by other methods for safety, environmental, or economic reasons. However, recognition that incineration is not always appropriate should not obscure the fact that it is an excellent way to dispose of most chemically contaminated biological waste.

CITED REFERENCES

1. S. S. Block and J. C. Netherton, "Infectious hospital wastes: their treatment and disposal," Chapter 36 in *Disinfection, Sterilization, and Preservation,* S. S. Block, ed., Lea & Febiger, Philadelphia, Pa., 1977, pp. 723–739.
2. J. J. Perkins. *Principles and Methods of Sterilization in Health Sciences,* 2nd ed., Charles C Thomas, Springfield, Ill., 1969.
3. S. D. Rubbo and J. F. Gardner, eds., *A Review of Sterilization and Disinfection,* Lloyd-Luke Ltd., London, 1965.
4. U.S. Nuclear Regulatory Commission, *Rules and Regulations,* 10 CFR Chapter 1.

GENERAL REFERENCE WORKS

W. E. Barkley and A. G. Wedum, "The hazard of infectious agents in microbiological laboratories," in *Disinfection, Sterilization, and Preservation,* S. S. Block, ed., Lea & Febiger, Philadelphia, Pa., 1977.

W. Horowitz, ed., "Disinfectants," Chapter 4 in *Official Methods of Analysis of the Association of Official Analytical Chemists,* 13th ed., Assoc. of Analytical Chemists, Washington, D.C., 1980, pp. 56–68.

G. B. Phillips, "Control of microbiological hazards in the laboratory," *Am. Ind. Hyg. Assoc. J. 30,* 170–176, 1969.

D. A. Shapton and R. G. Board, *Safety in Microbiology,* Technical Series #6, Society of Applied Bacteriology, Academic Press, New York, 1972.

U.S. Pharmacopeial Convention, Inc., "Sterilization," Section 1211 in *The United States Pharmacopeia,* 20th revision, U.S. Pharmacopeial Convention, Inc., Rockville, Md., 1980, pp. 1037–1039.

GOVERNMENT GUIDELINES

U.S. Department of Agriculture, "Disposal of diseased or otherwise adulterated carcasses and parts," in U.S. Code of Federal Regulations, Title 9, U.S. Government Printing Office, Washington, D.C., 1981.

U.S. Department of Health and Human Services, Centers for Disease Control, "Guidelines for hospital environmental control: cleaning, disinfection and sterilization of hospital equipment, in *Guidelines for the Prevention and Control of Nosocomial Infections*, CDC, Center for Infectious Diseases, Atlanta, Ga., February 1981.

U.S. Department of Health and Human Services, Centers for Disease Control, *Proposed Biosafety Guidelines for Microbiological and Biomedical Laboratories*, CDC, Office of Biosafety, Atlanta, Ga., 1980.

U.S. Department of Health and Human Services, Centers for Disease Control, *Disposal of Solid Wastes from Hospitals*, 2nd revision, CDC, Bureau of Epidemiology, Atlanta, Ga., June 1980.

U.S. Department of Health, Education, and Welfare, Public Health Service, National Institutes of Health, "Human genetic mutant cell repository minimum safety guidelines recommended for working with lymphoid and virus transformed human cell lines," in *The Human Genetic Mutant Cell Repository,* 4th ed., October 1977.

U.S. Department of Health, Education, and Welfare, Public Health Service, National Institutes of Health, National Cancer Institute, Office of Research Safety, and the Special Committee of Safety and Health Experts, *Laboratory Safety Monograph— Supplements 1 and 2 to the NIH Guidelines for Recombinant DNA Research*, July 1978 and January 1979.

U.S. Environmental Protection Agency. *Draft Manual for Infectious Waste Management*, SW-957, September 1982.

SPECIFIC PERTINENT REFERENCES

M. S. Barbeito and E. A. Brookey, Jr., "Microbiological hazard from the exhaust of a high-vacuum sterilizer," *Appl. Environ. Microbiol. 32*, 671–678, 1976.

M. S. Barbeito and G. G. Gremillion, "Microbiological safety evaluation of an industrial refuse incinerator," *Appl. Microbiol. 16*, 291–295, 1968.

M. S. Barbeito, L. A. Taylor, and R. W. Seiders, "Microbiological evaluation of a large-volume air incinerator," *Appl. Microbiol. 16*, 490–495, 1968.

H. Hess, L. Geller, and X. Buehlmann, "Ethylene oxide treatment of naturally contaminated materials," Chapter 15 in *Industrial Sterilization*, G. B. Phillips and W. S. Miller, eds., Duke University Press, Durham, N.C., 1973, pp. 283–296.

S. Kaye, "Disposing of laboratory disposables," *Lab. Mgmt. 16*, 37–44, 1978.

C. H. Lee, T. J. Montville, and A. J. Sinskey, "Comparison of the efficacy of steam sterilization indicators," *Appl. Environ. Microbiol. 37*, 1113–1117, 1979.

J. V. Rodricks, C. W. Hesseltine, and M. A. Mehlman, eds., *Mycotoxins in Human and Animal Health, Proceedings of the First International Conference*, Pathotox Publ., Inc., Park Forest S., Ill., 1977.

L. A. Taylor, M. S. Barbeito, and G. G. Gremillion, "Paraformaldehyde for surface sterilization and detoxification," *Appl. Microbiol. 17*, 614–618, 1969.

EPA RCRA Regulations Pertinent to Generators of Laboratory Waste

This appendix contains brief summaries of and copies of some EPA RCRA regulations that are pertinent to generators of hazardous waste. Included are the following:

40 CFR 261—Identification and Listing of Hazardous Waste
40 CFR 262—Standards Applicable to Generators of Hazardous Waste
40 CFR 263—Standards Applicable to Transporters of Hazardous Waste
40 CFR 265.316—Lab Packs

These regulations are those in effect 1 April 1983. Regulations are continually being changed by amendment, and readers must be alert to changes that occur after this date (see Appendix M).

This material is intended to provide a guide to these EPA RCRA regulations for small laboratories with a relatively few points of waste generation. Large organizations with many generation sites should refer to 40 CFR Parts 260–265 for the EPA RCRA regulations.

40 CFR 260.10—Definition of "Generator"

"Generator" means any person, by site, whose act or process produces hazardous waste identified or listed in Part 261.

"Person" means an individual, trust, firm, joint stock company, federal agency, corporation (including a government corporation), partnership, association, state, municipality, commission, political subdivision of a state, or any interstate body. (This definition should be construed to include colleges and universities.)

"Individual generation site" means the contiguous site at or on which one or more hazardous wastes are generated. An individual generation site, such as a large manufacturing plant, may have one or more sources of hazardous waste but is considered a single or individual generation site if the site or property is contiguous.

"On-site" means the same or geographically contiguous property that may be divided by public or private right-of-way, provided the entrance and exit between the properties is at a crossroads intersection and access is by crossing, as opposed to going along, the right-of-way. Noncontiguous properties owned by the same person but connected by a right-of-way that the person controls and to which the public does not have access are also considered on-site property.

Questions concerning the interpretation of these definitions should be addressed to the pertinent regional EPA office.

Region I: Maine, Vermont, New Hampshire, Massachusetts, Connecticut, and Rhode Island.
                John F. Kennedy Building, Boston, Mass. 02203; (617) 223-5777.

Region II: New York, New Jersey, Commonwealth of Puerto Rico, and the U.S. Virgin Islands.
                26 Federal Plaza, New York, N.Y. 10007; (212) 264-0504/5.

Region III: Pennsylvania, Delaware, Maryland, West Virginia, Virginia, and the District of Columbia.
                6th and Walnut Streets, Philadelphia, Pa. 19106; (215) 597-0980.

Region IV: Kentucky, Tennessee, North Carolina, Mississippi, Alabama, Georgia, South Carolina, and Florida.
                345 Courtland Street, N.E., Atlanta, Ga. 30365; (404) 881-3016.

Region V: Minnesota, Wisconsin, Illinois, Michigan, Indiana, and Ohio.
                230 South Dearborn Street, Chicago, Ill. 60604; (312) 886-6148.

Region VI: New Mexico, Oklahoma, Arkansas, Louisiana, and Texas.
               1201 Elm Street, First International Building, Dallas, Tex. 75270; (214) 767-2645.
Region VII: Nebraska, Kansas, Missouri, and Iowa.
               324 East 11th Street, Kansas City, Mo. 64106; (816) 374-3307.
Region VIII: Montana, Wyoming, North Dakota, South Dakota, Utah, and Colorado.
               1860 Lincoln Street, Denver, Colo. 80203; (303) 837-2221.
Region IX: California, Nevada, Arizona, Hawaii, Guam, American Samoa, and the Commonwealth of the Northern Mariana Islands.
               215 Fremont Street, San Francisco, Calif. 94105; (415) 556-4606.
Region X: Washington, Oregon, Idaho, and Alaska.
               1200 6th Avenue, Seattle, Wash. 98101; (206) 442-1260.

40 CFR 261—Identification and Listing of Hazardous Waste
Subpart A. General

261.2. The definition of a "solid waste" includes any solid, liquid, semisolid, or contained gaseous material that is "discarded" if it is abandoned (and not used, reused, reclaimed, or recycled) by being
1. Disposed of; or
2. Burned or incinerated, except where the material is being burned as a fuel for the purpose of recovering usable energy; or
3. Physically, chemically, or biologically treated (other than burned or incinerated) in lieu of or prior to being disposed of.

261.3. A solid waste is a *hazardous waste* if
1. It is not excluded from regulation under 261.4(b).
2. It exhibits any of the characteristics of a hazardous waste in Subpart C.
3. It is listed in Subpart D.

261.4. Wastes that are not classified as hazardous waste include
1. Domestic sewage.
2. Household waste.
3. Agricultural wastes that are returned to the soil as fertilizers.

4. Waste (e.g., ash) generated by combustion of fossil fuel.
5. Samples collected for the purpose of testing to determine characteristics or composition.

*261.5. Small Quantity Generator Exemption.**

1. If a generator produces, in a calendar month, less than 1000 kg of hazardous waste, that waste is not subject to the regulations under Parts 262–265 or to the notification requirements. The generator is responsible for determining whether a waste is an EPA-regulated hazardous waste (262.11). Note that this exemption could apply in one month and not in another. In the original publication of the EPA RCRA regulations in 1980, EPA stated its intention to lower the 1000-kg limit for this exemption to 100 kg in 2–5 years.

    Legislation to mandate this reduction, or to eliminate the exemption altogether, was introduced but not passed in the 1982 Congress. Laboratories that utilize this exemption should be alert for future amendments that may change the exemption level.

2. Small quantity generators are allowed indefinite and unregulated storage of up to 1000 kg of hazardous waste. However, if or when the 1000-kg limit is exceeded, the wastes become subject to full regulation, and the generator must then ship them off site within 90 days or obtain a storage permit.

3. Acutely hazardous waste [261.33(e)] generated in excess of 1 kg in a calendar month, or accumulated in excess of 1 kg, is subject to full regulation.

4. Hazardous waste that qualifies for the small quantity exemption must be treated or disposed of on site, or delivered by approved transport, with appropriate recordkeeping, to a disposal facility that is authorized by EPA or the state.

5. Hazardous waste that is exempt under this section remains exempt even if it is mixed with nonhazardous waste and the resulting mixture exceeds the quantity limit.

261.6. Special Requirements for Hazardous Waste That is Used, Reused, Recycled, or Reclaimed.

Hazardous waste, except sludges and wastes listed in Subpart D,

---

*The EPA is undertaking (1983) an analysis of alternative methods for regulating small quantity generators under RCRA. The analysis of small quantity generators will cover quantities of waste generated, current waste management practices (including treatment, use/reuse, recycling, and recovery practices), and waste management costs. The study is projected to require 2 years.

is not subject to regulation if it is being used, reused, recycled, or reclaimed or if it is being stored or treated prior to use, reuse, recycling, or reclamation.

261.7. Residues of Hazardous Waste in Empty Containers.

1. Hazardous waste remaining in an empty container is not subject to regulation.
2. A container or inner liner removed from a container that has held hazardous waste, except a waste that is a compressed gas or that is identified in 261.33, is empty if:
   a. All waste has been removed that can be removed using common practice and
   b. No more than 2.5 cm of residue remains on the bottom of the container or inner liner.
3. A container that has held a hazardous waste that is a compressed gas is empty when the pressure in the container approaches atmospheric.
4. A container or inner liner is empty if it has been triple-rinsed with a proper solvent or has been cleaned by another method shown in the literature or by tests to be equivalent to triple rinsing.

Subpart B. Criteria for Identifying the Characteristics of Hazardous Waste and for Listing Hazardous Waste. Subpart C. Characteristics of Hazardous Waste.

A solid waste is a hazardous waste if it exhibits any of the following characteristics:

261.21. Ignitability.

1. Liquids having a flash point below 60°C (140°F). Aqueous solutions containing less than 24% alcohol by volume are excluded.
2. Nonliquids liable to cause fires through friction, absorption of moisture, spontaneous chemical change, or retained heat are liable, when ignited, to burn so vigorously and persistently as to create a hazard.
3. Ignitable compressed gases.
4. Oxidizers as defined in 49 CFR 173.151: A substance such as a chlorate, permanganate, inorganic peroxide, or nitrate that yields oxygen readily to stimulate the combustion of organic matter.

261.22. Corrosivity.

1. Aqueous waste that has a pH $\leq$ 2.0 or $\geq$ 12.5.
2. Liquid wastes capable of corroding SAE 1020 steel at a rate greater than 0.635 cm (0.250 inch) per year.

261.23. Reactivity.

1. Wastes that readily undergo violent chemical change.

2. Wastes that react violently with or form potentially explosive mixtures with water.

3. Wastes that generate toxic fumes in a quantity sufficient to present a danger to human health or the environment when mixed with water or, in the case of cyanide- or sulfide-bearing wastes, when exposed to a pH between 2.0 and 12.5.

4. Wastes that explode when subjected to a strong initiating force, explode at normal temperatures and pressures, or fit within the DOT classifications of forbidden explosives (Appendix D, 49 CFR 173.51), Class A explosives (Appendix D, 49 CFR 173.53), or Class B explosives (Appendix D, 49 CFR 173.88).

261.24. Extraction Procedure (EP) Toxicity.

This characteristic identifies wastes from which certain toxic materials could be leached into groundwater supplies. This section includes the list of these materials and the extraction procedure.

Subpart D. Lists of Hazardous Wastes.

A solid waste is a hazardous waste if it is listed in this Subpart.

40 CFR 262—Standards Applicable to Generators of Hazardous Waste.

Subpart A. General.

262.11. Hazardous Waste Determination.

A person who generates a solid waste is responsible for determining whether it is a regulated hazardous waste.

262.12. EPA Identification Number.

A generator who treats, stores, disposes of, transports, or offers for transport hazardous waste must obtain an EPA Identification Number.

Subpart B. The Manifest.

262.20.

A generator who transports, or offers for transportation, hazardous waste for off-site treatment, storage, or disposal must prepare a manifest before transporting the waste off site. The generator must designate on the manifest a facility that has a permit to handle the waste.

262.21. Required Information.

Contains a list of all the information that must be included in the manifest.

262.23. Use of the Manifest.

Subpart C. Pretransport Requirements.

Requirements on packaging, labeling, marking, and placarding.

262.34. Accumulation Time.

Sets forth time limits for accumulation of hazardous waste on site.

EPA has proposed changes, "the Satellite Amendments," to this section (see 48 FR 118–121, 3 January 1983).

Subpart D. Recordkeeping and Reporting.

262.40. (as amended 28 January 1983).

Copies of each manifest, Biennial Report, Exception Report, and records of waste analyses must be kept for three years.

262.41. (as amended 28 January 1983).

A generator who ships hazardous waste off site must submit a biennial report to the Regional EPA Administrator by 1 March of each even-numbered year. The report must be on EPA Form 8700-13A and must cover the generator's activities during the previous calendar year. This section lists the specific information that must be included in the biennial report.

Appendix.

This Appendix to Section 262 was removed from CFR by amendment of 28 January 1983.

40 CFR 263. Standards Applicable to Transporters of Hazardous Waste. This section is included for information on dealing with contractors employed for off-site disposal of hazardous waste.

40 CFR 265 Subpart N—Landfills.

265.316. Lab Packs.

Describes the requirements for landfill disposal of small containers of hazardous waste in overpacked drums, i.e., lab packs.

(d) Incompatible wastes [260.10(a)]:

A hazardous waste that is unsuitable for:

1. Placement in a particular device or facility because it may cause corrosion or decay of containment materials (e.g., container inner liners or tank walls); or

2. Commingling with another waste or material under uncontrolled conditions because the commingling might produce heat or pressure, fire or explosion, violent reaction, toxic dusts, mists, fumes, or gases, or flammable fumes or gases.

(e) Reactive cyanide- or sulfide-bearing waste [261.23(a)(5)].

PART 261—IDENTIFICATION AND LISTING OF HAZARDOUS WASTE

Subpart A—General

AUTHORITY: Secs. 1006, 2002(a), 3001, and 3002 of the Solid Waste Disposal Act, as amended by the Resource Conservation and Recovery Act of 1976, as amended (42 U.S.C. 6905, 6912, 6921, and 6922).

SOURCE: 45 FR 33119, May 19, 1980, unless otherwise noted.

Subpart A—General

§ 261.1 Purpose and scope.

(a) This part identifies those solid wastes which are subject to regulation as hazardous wastes under Parts 262 through 265 and Parts 122 through 124 of this Chapter and which are subject to the notification requirements of section 3010 of RCRA. In this part:

(1) Subpart A defines the terms "solid waste" and "hazardous waste," identifies those wastes which are excluded from regulation under Parts 262 through 265 and 122 through 124 and establishes special management requirements for hazardous waste produced by small quantity generators and hazardous waste which is used, re-used, recycled or reclaimed.

(2) Subpart B sets forth the criteria used by EPA to identify characteristics of hazardous waste and to list particular hazardous wastes.

(3) Subpart C identifies characteristics of hazardous waste.

(4) Subpart D lists particular hazardous wastes.

(b) This part identifies only some of the materials which are hazardous wastes under sections 3007 and 7003 of RCRA. A material which is not a hazardous waste identified in this part is still a hazardous waste for purposes of those sections if:

(1) In the case of section 3007, EPA has reason to believe that the material may be a hazardous waste within the meaning of section 1004(5) of RCRA.

(2) In the case of section 7003, the statutory elements are established.

§ 261.2 Definition of solid waste.

(a) A solid waste is any garbage, refuse, sludge or any other waste material which is not excluded under § 261.4(a).

(b) An "other waste material" is any solid, liquid, semi-solid or contained gaseous material, resulting from industrial, commercial, mining or agricultural operations, or from community activities which:

(1) Is discarded or is being accumulated, stored or physically, chemically or biologically treated prior to being discarded; or

(2) Has served its original intended use and sometimes is discarded; or

(3) Is a manufacuring or mining by-product and sometimes is discarded.

(c) A material is "discarded" if it is abandoned (and not used, re-used, reclaimed or recycled) by being:

(1) Disposed of; or

(2) Burned or incinerated, except where the material is being burned as a fuel for the purpose of recovering usable energy; or

(3) Physically, chemically, or biologically treated (other than burned or incinerated) in lieu of or prior to being disposed of.

(d) A material is "disposed of" if it is discharged, deposited, injected, dumped, spilled, leaked or placed into or on any land or water so that such material or any constituent thereof may enter the environment or be emitted into the air or discharged into ground or surface waters.

(e) A "manufacturing or mining by-product" is a material that is not one of the primary products of a particular manufacturing or mining operation, is a secondary and incidental product of the particular operation and would not be solely and separately manufactured or mined by the particular manufacturing or mining operation. The term does not include an intermediate manufacturing or mining product which results from one of the steps in a manufacturing or mining process and is typically processed through the next step of the process within a short time.

§ 261.3 Definition of hazardous waste.

(a) A solid waste, as defined in § 261.2, is a hazardous waste if:

(1) It is not excluded from regulation as a hazardous waste under § 261.4(b); and

(2) It meets any of the following criteria:

(i) It exhibits any of the characteristics of hazardous waste identified in Subpart C.

(ii) It is listed in Subpart D and has not been excluded from the lists in Subpart D under §§ 260.20 and 260.22 of this chapter.

(iii) It is a mixture of a solid waste and a hazardous waste that is listed in Subpart D solely because it exhibits one or more of the characteristics of hazardous waste identified in Subpart C, unless the resultant mixture no longer exhibits any characteristic of hazardous waste identified in Subpart C.

(iv) It is a mixture of solid waste and one or more hazardous wastes listed in Subpart D and has not been excluded from this paragraph under §§ 260.20 and 260.22 of this chapter; however, the following mixtures of solid wastes and hazardous wastes listed in Subpart D are not hazardous wastes (except by application of paragraph (a)(2) (i) or (ii) of this section) if the generator can demonstrate that the mixture consists of wastewater the discharge of which is subject to regulation under either Section 402 or Section 307(b) of the Clean Water Act (including wastewater at facilities which have eliminated the discharge of wastewater) and:

(A) One or more of the following spent solvents listed in § 261.31—carbon tetrachloride, tetrachloroethylene, trichoroethylene—provided that the maximum total weekly usage of these solvents (other than the amounts that can be demonstrated not to be discharged to wastewater) divided by the average weekly flow of wastewater into the headworks of the facility's wastewater treatment or pretreatment system does not exceed 1 part per million; or

(B) One or more of the following spent solvents listed in § 261.31—methylene chloride, 1,1,1-trichloroethane, chlorobenzene, o-dichlorobenzene, cresols, cresylic acid, nitrobenzene, toluene, methyl ethyl ketone, carbon disulfide, isobutanol, pyridine, spent chlorofluorocarbon solvents—provided that the maximum total weekly usage of these solvents (other than the amounts that can be demonstrated not to be discharged to wastewater) divided by the average weekly flow of wastewater into the headworks of the facility's wastewater treatment or pretreatment system does not exceed 25 parts per million; or

(C) One of the following wastes listed in § 261.32—heat exchanger bundle cleaning sludge from the petroleum refining industry (EPA Hazardous Waste No. K050); or

(D) A discarded commercial chemical product, or chemical intermediate listed in § 261.33, arising from *de minimis* losses of these materials from manufacturing operations in which these materials are used as raw materials or are produced in the manufacturing process. For purposes of this subparagraph, *"de minimis"* losses include those from normal material handling operations (e.g. spills from the unloading or transfer of materials from bins or other containers, leaks from pipes, valves or other devices used to transfer materials); minor leaks of process equipment, storage tanks or containers; leaks from well-maintained pump packings and seals; sample purgings; relief device discharges; discharges from safety showers and rinsing and cleaning of personal safety equipment; and rinsate from empty containers or from containers that are rendered empty by that rinsing; or

(E) Wastewater resulting from laboratory operations containing toxic (T) wastes listed in Subpart D, provided that the annualized average flow of laboratory wastewater does not exceed one percent of total wastewater flow into the headworks of the facility's wastewater treatment or pre-treatment system, or provided the wastes, combined annualized average concentration does not exceed one part per million in the headworks of the facility's wastewater treatment or pre-treatment facility. Toxic (T) wastes used in laboratories that are demonstrated not to be discharged to wastewater are not to be included in this calculation.

(b) A solid waste which is not excluded from regulation under paragraph (a)(1) of this section becomes a hazardous waste when any of the following events occur:

(1) In the case of a waste listed in Subpart D, when the waste first meets the listing description set forth in Subpart D.

(2) In the case of a mixture of solid waste and one or more listed hazardous wastes, when a hazardous waste listed in Subpart D is first added to the solid waste.

(3) In the case of any other waste (including a waste mixture), when the waste exhibits any of the characteristics identified in Subpart C.

(c) Unless and until it meets the criteria of paragraph (d):

(1) A hazardous waste will remain a hazardous waste.

(2) Any solid waste generated from the treatment, storage or disposal of a hazardous waste, including any sludge, spill residue, ash, emission control dust or leachate (but not including precipitation run-off), is a hazardous waste.

(d) Any solid waste described in paragraph (c) of this section is not a hazardous waste if it meets the following criteria:

(1) In the case of any solid waste, it does not exhibit any of the characteristics of hazardous waste identified in Subpart C.

(2) In the case of a waste which is a listed waste under Subpart D, contains a waste listed under Subpart D or is derived from a waste listed in Subpart D, it also has been excluded from paragraph (c) under §§ 260.20 and 260.22 of this Chapter.

[45 FR 33119, May 19, 1980, as amended at 46 FR 56588, Nov. 11, 1981]

§ 261.4 Exclusions.

(a) *Materials which are not solid wastes.* The following materials are not solid wastes for the purpose of this part:

(1)(i) Domestic sewage; and

(ii) Any mixture of domestic sewage and other wastes that passes through a sewer system to a publicly-owned treatment works for treatment. "Domestic sewage" means untreated sanitary wastes that pass through a sewer system.

(2) Industrial wastewater discharges that are point source discharges subject to regulation under Section 402 of the Clean Water Act, as amended.

[*Comment:* This exclusion applies only to the actual point source discharge. It does not exclude industrial wastewaters while they are being collected, stored or treated before discharge, nor does it exclude sludges that are generated by industrial wastewater treatment.]

(3) Irrigation return flows.

(4) Source, special nuclear or by-product material as defined by the Atomic Energy Act of 1954, as amended, 42 U.S.C. 2011 *et seq.*

(5) Materials subjected to in-situ mining techniques which are not removed from the ground as part of the extraction process.

(b) *Solid wastes which are not hazardous wastes.* The following solid wastes are not hazardous wastes:

(1) Household waste, including household waste that has been collected, transported, stored, treated, disposed, recovered (e.g., refuse-derived fuel) or reused. "Household waste" means any waste material (including garbage, trash and sanitary wastes in septic tanks) derived from households (including single and multiple residences, hotels and motels.)

(2) Solid wastes generated by any of the following and which are returned to the soils as fertilizers:

(i) The growing and harvesting of agricultural crops.

(ii) The raising of animals, including animal manures.

(3) Mining overburden returned to the mine site.

(4) Fly ash waste, bottom ash waste, slag waste, and flue gas emission control waste generated primarily from the combustion of coal or other fossil fuels.

(5) Drilling fluids, produced waters, and other wastes associated with the exploration, development, or production of crude oil, natural gas or geothermal energy.

(6)(i) Wastes which fail the test for the characteristic of EP toxicity because chromium is present or are listed in Subpart D due to the presence of chromium, which do not fail the test for the characteristic of EP toxicity for any other constituent or are not listed due to the presence of any other constituent, and which do not fail the text for any other characteristic, if it is shown by a waste generator or by waste generators that:

(A) The chromium in the waste is exclusively (or nearly exclusively) trivalent chromium; and

(B) The waste is generated from an industrial process which uses trivalent chromium exlcusively (or nearly exclusively) and the process does not generate hexavalent chromium; and

(C) The waste is typically and frequently managed in non-oxidizing environments.

(ii) Specific wastes which meet the standard in paragraphs (b)(6)(i)(A), (B) and (C) (so long as they do not fail the test for the charactristic of EP toxicity, and do not fail the test for any other characteristic) are:

(A) Chrome (blue) trimmings generated by the following subcategories of the leather tanning and finishing industry; hair pulp/chrome tan/retan/ wet finish; hair save/chrome tan/ retan/wet finish; retan/wet finish; no beamhouse; through-the-blue; and shearling.

(B) Chrome (blue) shavings generated by the following subcategories of the leather tanning and finishing industry: hair pulp/chrome tan/retan/ wet finish; hair save/chrome tan/ retan/wet finish; retan/wet finish; no beamhouse; through-the-blue; and shearling.

(C) Buffing dust generated by the following subcategories of the leather tanning and finishing industry; hair pulp/chrome tan/retan/wet finish; hair save/chrome tan/retan/wet finish; retan/wet finish; no beamhouse; through-the-blue.

(D) Sewer screenings generated by the following subcategories of the leather tanning and finishing industry: hair pulp/crome tan/retan/wet finish; hair save/chrome tan/retan/ wet finish; retan/wet finish; no beamhouse; through-the-blue; and shearling.

(E) Wastewater treatment sludges generated by the following subcategories of the leather tanning and finishing industry: hair pulp/chrome tan/ retan/wet finish; hair save/chrome tan/retan/wet finish; retan/wet finish; no beamhouse; through-the-blue; and shearling.

(F) Wastewater treatment sludes generated by the following subcategories of the leather tanning and finishing industry: hair pulp/chrome tan/retan/wet finish; hair save/chrome-tan/retan/wet finish; and through-the-blue.

(G) Waste scrap leather from the leather tanning industry, the shoe manufacturing industry, and other leather product manufacturing industries.

(H) Wastewater treatment sludges from the production of TiO_2 pigment using chromium-bearing ores by the chloride process.

(7) Solid waste from the extraction, beneficiation and processing of ores and minerals (including coal), including phosphate rock and overburden from the mining of uranium ore.

(8) Cement kiln dust waste.

(9) Solid waste which consists of discarded wood or wood products which fails the test for the characteristic of EP toxicity and which is not a hazardous waste for any other reason if the waste is generated by persons who utilize the arsenical-treated wood and wood products for these materials' intended end use.

(c) Hazardous wastes which are exempted from certain regulations. A hazardous waste which is generated in a product or raw material storage tank, a product or raw material transport vehicle or vessel, a product or raw material pipeline, or in a manufacturing process unit or an associated non-waste-treatment-manufacturing unit, is not subject to regulation under Parts 262 through 265 and Parts 122 through 124 of this chapter or to the notification requirements of Section 3010 of RCRA until it exits the unit in which it was generated, unless the unit is a surface impoundment, or unless the hazardous waste remains in the unit more than 90 days after the unit ceases to be operated for manufacturing, or for storage or transportation of product or raw materials.

(d) *Samples.* (1) Except as provided in paragraph (d)(2) of this section, a sample of solid waste or a sample of water, soil, or air, which is collected for the sole purpose of testing to determine its characteristics or composition, is not subject to any requirements of this part or Parts 262 through 267 or Part 122 or Part 124 of this chapter or to the notification requirements of Section 3010 of RCRA, when:

(i) The sample is being transported to a laboratory for the purpose of testing; or

(ii) The sample is being transported back to the sample collector after testing; or

(iii) The sample is being stored by the sample collector before transport to a laboratory for testing; or

(iv) The sample is being stored in a laboratory before testing; or

(v) The sample is being stored in a laboratory after testing but before it is returned to the sample collector; or

(vi) The sample is being stored temporarily in the laboratory after testing for a specific purpose (for example, until conclusion of a court case or enforcement action where further testing of the sample may be necessary).

(2) In order to qualify for the exemption in paragraphs (d)(1) (i) and (ii) of this section, a sample collector shipping samples to a laboratory and a laboratory returning samples to a sample collector must:

(i) Comply with U.S. Department of Transportation (DOT), U.S. Postal Service (USPS), or any other applicable shipping requirements; or

(ii) Comply with the following requirements if the sample collector determines that DOT, USPS, or other shipping requirements do not apply to the shipment of the sample:

(A) Assure that the following information accompanies the sample:

(*1*) The sample collector's name, mailing address, and telephone number;

(*2*) The laboratory's name, mailing address, and telephone number;

(*3*) The quantity of the sample;

(*4*) The date of shipment; and

(*5*) A description of the sample.

(B) Package the sample so that it does not leak, spill, or vaporize from its packaging.

(3) This exemption does not apply if the laboratory determines that the waste is hazardous but the laboratory is no longer meeting any of the conditions stated in paragraph (d)(1) of this section.

[45 FR 33119, May 19, 1980, as amended at 45 FR 72037, Oct. 30, 1980; 45 FR 76620,

Nov. 19, 1980; 45 FR 78531, Nov. 25, 1980; 45 FR 80287, Dec. 4, 1980; 46 FR 27476, May 20, 1981; 46 FR 47429, Sept. 25, 1981]

§ 261.5 Special requirements for hazardous waste generated by small quantity generators.

(a) A generator is a small quantity generator in a calendar month if he generates less than 1000 kilograms of hazardous waste in that month.

(b) Except for those wastes identified in paragraphs (e) and (f) of this section, a small quantity generator's hazardous wastes are not subject to regulation under Parts 262 through 265 and Parts 122 and 124 of this chapter, and the notification requirements of Section 3010 of RCRA, provided the generator complies with the requirements of paragraph (g) of this section.

(c) Hazardous waste that is beneficially used or re-used or legitimately recycled or reclaimed and that is excluded from regulation by § 261.6(a) is not included in the quantity determinations of this section, and is not subject to any requirements of this section. Hazardous waste that is subject to the special requirements of § 261.6(b) is included in the quantity determinations of this section and is subject to the requirements of this section.

(d) In determining the quantity of hazardous waste he generates, a generator need not include:

(1) His hazardous waste when it is removed from on-site storage; or

(2) Hazardous waste produced by on-site treatment of his hazardous waste.

(e) If a small quantity generator generates acutely hazardous waste in a calendar month in quantities greater than set forth below, all quantities of that acutely hazardous waste are subject to regulation under Parts 262 through 265 and Parts 122 and 124 of this chapter, and the notification requirements of Section 3010 of RCRA:

(1) A total of one kilogram of commercial chemical products and manufacturing chemical intermediates having the generic names listed in § 261.33(e), and off-specification commercial chemical products and manufacturing chemical intermediates which, if they met specifications, would have the generic names listed in § 261.33(e).

(2) A total of 100 kilograms of any residue or contaminated soil, water or other debris resulting from the cleanup of a spill, into or on any land or water, of any commercial chemical products or manufacturing chemical intermediates having the generic names listed in § 261.33(e), or any residue or contaminated soil, water or other debris resulting from the cleanup of a spill, into or on any land or water, of any off-specification commercial chemical products or manufacturing chemical intermediates which, if they met specifications, would have the generic names listed in § 261.33(e).

(f) A small quantity generator may accumulate hazardous waste on-site. If he accumulates at any time more than a total of 1000 kilograms of his hazardous waste, or his acutely hazardous wastes in quantities greater than set forth in paragraph (e)(1) or (e)(2) of this section, all of those accumulated wastes for which the accumulation limit was exceeded are subject to regulation under Parts 262 through 265 and Parts 122 and 124 of this chapter, and the notification requirements of Section 3010 of RCRA. The time period of § 262.34 for accumulation of wastes on-site begins for a small quantity generator when the accumulated wastes exceed the applicable exclusion level.

(g) In order for hazardous waste generated by a small quantity generator to be excluded from full regulation under this section, the generator must:

(1) Comply with § 262.11 of this chapter;

(2) If he stores his hazardous waste on-site, store it in compliance with the requirements of paragraph (f) of this section; and

(3) Either treat or dispose of his hazardous waste in an on-site facility, or ensure delivery to an off-site storage, treatment or disposal facility, either of which is:

(i) Permitted under Part 122 of this chapter;

(ii) In interim status under Parts 122 and 265 of this chapter;

(iii) Authorized to manage hazardous waste by a State with a hazardous waste management program approved under Part 123 of this chapter;

(iv) Permitted, licensed or registered by a State to manage municipal or industrial solid waste; or

(v) A facility which:

(A) Beneficially uses or re-uses, or legitimately recycles or reclaims his waste; or

(B) Treats his waste prior to beneficial use or re-use, or legitimate recycling or reclamation.

(h) Hazardous waste subject to the reduced requirements of this section may be mixed with non-hazardous waste and remain subject to these reduced requirements even though the resultant mixture exceeds the quantity limitations identified in this section, unless the mixture meets any of the characteristics of hazardous wastes identified in Subpart C.

(i) If a small quantity generator mixes a solid waste with a hazardous waste that exceeds a quantity exclusion level of this section, the mixture is subject to full regulation.

(Secs. 1006, 2002(a) and 3002 of the Solid Waste Disposal Act, as amended by the Resource Conservation and Recovery Act of 1976 (RCRA), as amended, (42 U.S.C. 6905, 6912(a), and 6922))

[45 FR 76623, Nov. 19, 1980, as amended at 46 FR 27476, May 20, 1981; 46 FR 34587, July 2, 1981]

§ 261.6 Special requirements for hazardous waste which is used, re-used, recycled or reclaimed.

(a) Except as otherwise provided in paragraph (b) of this section, a hazardous waste which meets any of the following criteria is not subject to regulation under Parts 262 through 265 or Parts 122 through 124 of this Chapter and is not subject to the notification requirements of Section 3010 of RCRA until such time as the Administrator promulgates regulations to the contrary:

(1) It is being beneficially used or reused or legitimately recycled or reclaimed.

(2) It is being accumulated, stored or physically, chemically or biologically treated prior to beneficial use or re-use or legitimate recycling or reclamation.

(3) It is one of the following materials being used, reused, recycled or reclaimed in the specified manner:

(i) Spent pickle liquor which is reused in wastewater treatment at a facility holding a National Pollutant Discharge Elimination System (NPDES) permit, or which is being accumulated, stored, or physically, chemically or biologically treated before such reuse.

(b) Except for those wastes listed in paragraph (a)(3) of this section, a hazardous waste which is a sludge, or which is listed in Subpart D, or which contains one or more hazardous wastes listed in Subpart D; and which is transported or stored prior to being used, reused, recycled or reclaimed is subject to the following requirements with respect to such transportation or storage:

(1) Notification requirements under Section 3010 RCRA.

(2) Part 262 of this Chapter.

(3) Part 263 of this Chapter.

(4) Subparts A, B, C, D and E of Part 264 of this Chapter.

(5) Subparts A, B, C, D, E, G, H, I, J and L of Part 265 of this Chapter.

(6) Parts 122 and 124 of this Chapter, with respect to storage facilities.

[45 FR 33119, May 19, 1980, as amended at 46 FR 44973, Sept. 8, 1981]

§ 261.7 Residues of hazardous waste in empty containers.

(a)(1) Any hazardous waste remaining in either (i) an empty container or (ii) an inner liner removed from an empty container, as defined in paragraph (b) of this section, is not subject to regulation under Parts 261 through 265, or Part 122 or 124 of this chapter or to the notification requirements of Section 3010 of RCRA.

(2) Any hazardous waste in either (i) a container that is not empty or (ii) an inner liner removed from a container that is not empty, as defined in paragraph (b) of this section, is subject to regulation under Parts 261 through 265, and Parts 122 and 124 of this chapter and to the notification requirements of Section 3010 of RCRA.

(b)(1) A container or an inner liner removed from a container that has held any hazardous waste, except a waste that is a compressed gas or that is identified in § 261.33(c) of this chapter, is empty if:

(i) all wastes have been removed that can be removed using the practices commonly employed to remove materials from that type of container, *e.g.,* pouring, pumping, and aspirating, and

(ii) no more than 2.5 centimeters (one inch) of residue remain on the bottom of the container or inner liner.

(2) A container that has held a hazardous waste that is a compressed gas is empty when the pressure in the container approaches atmospheric.

(3) A container or an inner liner removed from a container that has held a hazardous waste identified in § 261.33(c) of this chapter is empty if:

(i) the container or inner liner has been triple rinsed using a solvent capable of removing the commercial chemical product or manufacturing chemical intermediate;

(ii) the container or inner liner has been cleaned by another method that has been shown in the scientific literature, or by tests conducted by the generator, to achieve equivalent removal; or

(iii) in the case of a container, the inner liner that prevented contact of the commercial chemical product or manufacturing chemical intermediate with the container, has been removed.

[45 FR 78529, Nov.25, 1980]

Subpart B—Criteria for Identifying the Characteristics of Hazardous Waste and for Listing Hazardous Waste

§ 261.10 Criteria for identifying the characteristics of hazardous waste.

(a) The Administrator shall identify and define a characteristic of hazardous waste in Subpart C only upon determining that:

(1) A solid waste that exhibits the characteristic may:

(i) Cause, or significantly contribute to, an increase in mortality or an increase in serious irreversible, or incapacitating reversible, illness; or

(ii) Pose a substantial present or potential hazard to human health or the environment when it is improperly

treated, stored, transported, disposed of or otherwise managed; and

(2) The characteristic can be:

(i) Measured by an available standardized test method which is reasonably within the capability of generators of solid waste or private sector laboratories that are available to serve generators of solid waste; or

(ii) Reasonably detected by generators of solid waste through their knowledge of their waste.

§ 261.11 Criteria for listing hazardous waste.

(a) The Administrator shall list a solid waste as a hazardous waste only upon determining that the solid waste meets one of the following criteria:

(1) It exhibits any of the characteristics of hazardous waste identified in Subpart C.

(2) It has been found to be fatal to humans in low doses or, in the absence of data on human toxicity, it has been shown in studies to have an oral LD 50 toxicity (rat) of less than 50 milligrams per kilogram, an inhalation LC 50 toxicity (rat) of less than 2 milligrams per liter, or a dermal LD 50 toxicity (rabbit) of less than 200 milligrams per kilogram or is otherwise capable of causing or significantly contributing to an increase in serious irreversible, or incapacitating reversible, illness. (Waste listed in accordance with these criteria will be designated Acute Hazardous Waste.)

(3) It contains any of the toxic constituents listed in Appendix VIII unless, after considering any of the following factors, the Administrator concludes that the waste is not capable of posing a substantial present or potential hazard to human health or the environment when improperly treated, stored, transported or disposed of, or otherwise managed:

(i) The nature of the toxicity presented by the constituent.

(ii) The concentration of the constituent in the waste.

(iii) The potential of the constituent or any toxic degradation product of the constituent to migrate from the waste into the environment under the types of improper management considered in paragraph (a)(3)(vii) of this section.

(iv) The persistence of the constituent or any toxic degradation product of the constituent.

(v) The potential for the constituent or any toxic degradation product of the constituent to degrade into non-harmful constituents and the rate of degradation.

(vi) The degree to which the constituent or any degradation product of the constituent bioaccumulates in ecosystems.

(vii) The plausible types of improper management to which the waste could be subjected.

(viii) The quantities of the waste generated at individual generation sites or on a regional or national basis.

(ix) The nature and severity of the human health and environmental damage that has occurred as a result of the improper management of wastes containing the constituent.

(x) Action taken by other governmental agencies or regulatory programs based on the health or environmental hazard posed by the waste or waste constituent.

(xi) Such other factors as may be appropriate.

Substances will be listed on Appendix VIII only if they have been shown in scientific studies to have toxic, carcinogenic, mutagenic or teratogenic effects on humans or other life forms. (Wastes listed in accordance with these criteria will be designated Toxic wastes.)

(b) The Administrator may list classes or types of solid waste as hazardous waste if he has reason to believe that individual wastes, within the class or type of waste, typically or frequently are hazardous under the definition of hazardous waste found in Section 1004(5) of the Act.

(c) The Administrator will use the criteria for listing specified in this section to establish the exclusion limits referred to in § 261.5(c).

Subpart C—Characteristics of Hazardous Waste

§ 261.20 General.

(a) A solid waste, as defined in § 261.2, which is not excluded from regulation as a hazardous waste under § 261.4(b), is a hazardous waste if it exhibits any of the characteristics identified in this Subpart.

[Comment: § 262.11 of this Chapter sets forth the generator's responsibility to determine whether his waste exhibits one or more of the characteristics identified in this Subpart]

(b) A hazardous waste which is identified by a characteristic in this subpart, but is not listed as a hazardous waste in Subpart D, is assigned the EPA Hazardous Waste Number set forth in the respective characteristic in this Subpart. This number must be used in complying with the notification requirements of Section 3010 of the Act and certain recordkeeping and reporting requirements under Parts 262 through 265 and Part 122 of this Chapter.

(c) For purposes of this Subpart, the Administrator will consider a sample obtained using any of the applicable sampling methods specified in Appendix I to be a representative sample within the meaning of Part 260 of this Chapter.

[Comment: Since the Appendix I sampling methods are not being formally adopted by the Administrator, a person who desires to employ an alternative sampling method is not required to demonstrate the equivalency of his method under the procedures set forth in §§ 260.20 and 260.21.]

§ 261.21 Characteristic of ignitability.

(a) A solid waste exhibits the characteristic of ignitability if a representative sample of the waste has any of the following properties:

(1) It is a liquid, other than an aqueous solution containing less than 24 percent alcohol by volume and has flash point less than 60°C (140°F), as determined by a Pensky-Martens Closed Cup Tester, using the test method specified in ASTM Standard D-93-79 or D-93-80 (incorporated by reference, see § 260.11), or a Setaflash Closed Cup Tester, using the test

method specified in ASTM Standard D–3278–78 (incorporated by reference, see § 260.11), or as determined by an equivalent test method approved by the Administrator under procedures set forth in §§ 260.20 and 260.21.

(2) It is not a liquid and is capable, under standard temperature and pressure, of causing fire through friction, absorption of moisture or spontaneous chemical changes and, when ignited, burns so vigorously and persistently that is creates a hazard.

(3) It is an ignitable compressed gas as defined in 49 CFR 173.300 and as determined by the test methods described in that regulation or equivalent test methods approved by the Administrator under §§ 260.20 and 260.21.

(4) It is an oxidizer as defined in 49 CFR 173.151.

(b) A solid waste that exhibits the characteristic of ignitability, but is not listed as a hazardous waste in Subpart D, has the EPA Hazardous Waste Number of D001.

[45 FR 33119, May 19, 1980, as amended at 46 FR 35247, July 7, 1981]

§ 261.22 Characteristic of corrosivity.

(a) A solid waste exhibits the characteristic of corrosivity if a representative sample of the waste has either of the following properties:

(1) It is aqueous and has a pH less than or equal to 2 or greater than or equal to 12.5, as determined by a pH meter using either an EPA test method or an equivalent test method approved by the Administrator under the procedures set forth in §§ 260.20 and 260.21. The EPA test method for pH is specified as Method 5.2 in "Test Methods for the Evaluation of Solid Waste, Physical/Chemical Methods" (incorporated by reference, see § 260.11).

(2) It is a liquid and corrodes steel (SAE 1020) at a rate greater than 6.35 mm (0.250 inch) per year at a test temperature of 55°C (130°F) as determined by the test method specified in NACE (National Association of Corrosion Engineers) Standard TM-01-69 as standardized in "Test Methods for the Evaluation of Solid Waste, Physical/Chemical Methods" (incorporated by reference, see § 260.11) or an equivalent test method approved by the Ad-

ministrator under the procedures set forth in §§ 260.20 and 260.21.

(b) A solid waste that exhibits the characteristic of corrosivity, but is not listed as a hazardous waste in Subpart D, has the EPA Hazardous Waste Number of D002.

[45 FR 33119, May 19, 1980, as amended at 46 FR 35247, July 7, 1981]

§ 261.23 Characteristic of reactivity.

(a) A solid waste exhibits the characteristic of reactivity if a representative sample of the waste has *any* of the following properties:

(1) It is normally unstable and readily undergoes violent change without detonating.

(2) It reacts violently with water.

(3) It forms potentially explosive mixtures with water.

(4) When mixed with water, it generates toxic gases, vapors or fumes in a quantity sufficient to present a danger to human health or the environment.

(5) It is a cyanide or sulfide bearing waste which, when exposed to pH conditions between 2 and 12.5, can generate toxic gases, vapors or fumes in a quantity sufficient to present a danger to human health or the environment.

(6) It is capable of detonation or explosive reaction if it is subjected to a strong initiating source or if heated under confinement.

(7) It is readily capable of detonation or explosive decomposition or reaction at standard temperature and pressure.

(8) It is a forbidden explosive as defined in 49 CFR 173.51, or a Class A explosive as defined in 49 CFR 173.53 or a Class B explosive as defined in 49 CFR 173.88.

(b) A solid waste that exhibits the characteristic of reactivity, but is not listed as a hazardous waste in Subpart D, has the EPA Hazardous Waste Number of D003.

§ 261.24 Characteristic of EP toxicity.

(a) A solid waste exhibits the characteristic of EP toxicity if, using the test methods described in Appendix II or equivalent methods approved by the Administrator under the procedures set forth in §§ 260.20 and 260.21, the extract from a representative sample of the waste contains any of the contaminants listed in Table I at a concentration equal to or greater than the respective value given in that Table. Where the waste contains less than 0.5 percent filterable solids, the waste itself, after filtering, is considered to be the extract for the purposes of this section.

(b) A solid waste that exhibits the characteristic of EP toxicity, but is not listed as a hazardous waste in Subpart D, has the EPA Hazardous Waste Number specified in Table I which corresponds to the toxic contaminant causing it to be hazardous.

TABLE I.—MAXIMUM CONCENTRATION OF CONTAMINANTS FOR CHARACTERISTIC OF EP TOXICITY

EPA hazardous waste number	Contaminant	Maximum concentration (milligrams per liter)
D004	Arsenic	5.0
D005	Barium	100.0
D006	Cadmium	1.0
D007	Chromium	5.0
D008	Lead	5.0
D009	Mercury	0.2
D010	Selenium	1.0
D011	Silver	5.0
D012	Endrin (1,2,3,4,10,10-hexachloro-1,7-epoxy-1,4,4a,5,6,7,8,8a-octahydro-1,4-endo, endo-5,8-dimethano-naphthalene.	0.02
D013	Lindane (1,2,3,4,5,6-hexa- chlorocyclohexane, gamma isomer.	0.4
D014	Methoxychlor (1,1,1-Trichloro-2,2-bis [p-methoxyphenyl]ethane).	10.0
D015	Toxaphene ($C_{10}H_{10}Cl_8$, Technical chlorinated camphene, 67–69 percent chlorine).	0.5
D016	2,4-D, (2,4-Dichlorophenoxyacetic acid).	10.0
D017	2,4,5-TP Silvex (2,4,5-Trichlorophenoxypropionic acid).	1.0

Subpart D—Lists of Hazardous Wastes

§ 261.30 General.

(a) A solid waste is a hazardous waste if it is listed in this Subpart, unless it has been excluded from this list under §§ 260.20 and 260.22.

(b) The Administrator will indicate his basis for listing the classes or types of wastes listed in this Subpart by employing one or more of the following Hazard Codes:

Ignitable Waste	(I)
Corrosive Waste	(C)
Reactive Waste	(R)
EP Toxic Waste	(E)
Acute Hazardous Waste	(H)
Toxic Waste	(T)

Appendix VII identifies the constituent which caused the Administrator to list the waste as an EP Toxic Waste (E) or Toxic Waste (T) in §§ 261.31 and 261.32.

(c) Each hazardous waste listed in this Subpart is assigned an EPA Hazardous Waste Number which precedes the name of the waste. This number must be used in complying with the notification requirements of Section 3010 of the Act and certain recordkeeping and reporting requirements under Parts 262 through 265 and Part 122 of this Chapter.

(d) The following hazardous wastes listed in § 261.31 or § 261.32 are subject to the exclusion limits for acutely hazardous wastes established in § 261.5: [Reserved]

[45 FR 33119, May 19, 1980, as amended at 45 FR 74892, Nov. 12, 1980]

§ 261.31   Hazardous wastes from non-specific sources.

Industry and EPA hazardous waste No.	Hazardous waste	Hazard code
Generic:		
F001	The following spent halogenated solvents used in degreasing: tetrachloroethylene, trichloroethylene, methylene chloride, 1,1,1-trichloroethane, carbon tetrachloride, and chlorinated fluorocarbons; and sludges from the recovery of these solvents in degreasing operations.	(T)
F002	The following spent halogenated solvents: tetrachloroethylene, methylene chloride, trichloroethylene, 1,1,1-trichloroethane, chlorobenzene, 1,1,2-trichloro-1,2,2-trifluoroethane, ortho-dichlorobenzene, and trichlorofluoromethane; and the still bottoms from the recovery of these solvents.	(T)
F003	The following spent non-halogenated solvents: xylene, acetone, ethyl acetate, ethyl benzene, ethyl ether, methyl isobutyl ketone, n-butyl alcohol, cyclohexanone, and methanol; and the still bottoms from the recovery of these solvents.	(I)
F004	The following spent non-halogenated solvents: cresols and cresylic acid, and nitrobenzene; and the still bottoms from the recovery of these solvents.	(T)
F005	The following spent non-halogenated solvents: toluene, methyl ethyl ketone, carbon disulfide, isobutanol, and pyridine; and the still bottoms from the recovery of these solvents.	(I, T)
F006	Wastewater treatment sludges from electroplating operations except from the following processes: (1) sulfuric acid anodizing of aluminum; (2) tin plating on carbon steel; (3) zinc plating (segregated basis) on carbon steel; (4) aluminum or zinc-aluminum plating on carbon steel; (5) cleaning/stripping associated with tin, zinc and aluminum plating on carbon steel; and (6) chemical etching and milling of aluminum.	(T)
F019	Wastewater treatment sludges from the chemical conversion coating of aluminum	(T)
F007	Spent cyanide plating bath solutions from electroplating operations (except for precious metals electroplating spent cyanide plating bath solutions).	(R, T)
F008	Plating bath sludges from the bottom of plating baths from electroplating operations where cyanides are used in the process (except for precious metals electroplating plating bath sludges).	(R, T)
F009	Spent stripping and cleaning bath solutions from electroplating operations where cyanides are used in the process (except for precious metals electroplating spent stripping and cleaning bath solutions).	(R, T)
F010	Quenching bath sludge from oil baths from metal heat treating operations where cyanides are used in the process (except for precious metals heat-treating quenching bath sludges).	(R, T)
F011	Spent cyanide solutions from salt bath pot cleaning from metal heat treating operations (except for precious metals heat treating spent cyanide solutions from salt bath pot cleaning).	(R, T)
F012	Quenching wastewater treatment sludges from metal heat treating operations where cyanides are used in the process (except for precious metals heat treating quenching wastewater treatment sludges).	(T)

[46 FR 4617, Jan. 16, 1981, as amended at 46 FR 27477, May 20, 1981]

§ 261.32 Hazardous wastes from specific sources.

Industry and EPA hazardous waste No.	Hazardous waste	Hazard code
Wood preservation: K001	Bottom sediment sludge from the treatment of wastewaters from wood preserving processes that use creosote and/or pentachlorophenol.	(T)
Inorganic pigments:		
K002	Wastewater treatment sludge from the production of chrome yellow and orange pigments.	(T)
K003	Wastewater treatment sludge from the production of molybdate orange pigments	(T)
K004	Wastewater treatment sludge from the production of zinc yellow pigments.................	(T)
K005	Wastewater treatment sludge from the production of chrome green pigments	(T)
K006	Wastewater treatment sludge from the production of chrome oxide green pigments (anhydrous and hydrated).	(T)
K007	Wastewater treatment sludge from the production of iron blue pigments	(T)
K008	Oven residue from the production of chrome oxide green pigments.............................	(T)
Organic chemicals:		
K009	Distillation bottoms from the production of acetaldehyde from ethylene	(T)
K010	Distillation side cuts from the production of acetaldehyde from ethylene.....................	(T)
K011	Bottom stream from the wastewater stripper in the production of acrylonitrile...............	(R, T)
K013	Bottom stream from the acetonitrile column in the production of acrylonitrile................	(R, T)
K014	Bottoms from the acetonitrile purification column in the production of acrylonitrile	(T)
K015	Still bottoms from the distillation of benzyl chloride..	(T)
K016	Heavy ends or distillation residues from the production of carbon tetrachloride............	(T)
K017	Heavy ends (still bottoms) from the purification column in the production of epichlorohydrin.	(T)
K018	Heavy ends from the fractionation column in ethyl chloride production........................	(T)
K019	Heavy ends from the distillation of ethylene dichloride in ethylene dichloride production.	(T)
K020	Heavy ends from the distillation of vinyl chloride in vinyl chloride monomer production.	(T)
K021	Aqueous spent antimony catalyst waste from fluoromethanes production	(T)
K022	Distillation bottom tars from the production of phenol/acetone from cumene.................	(T)
K023	Distillation light ends from the production of phthalic anhydride from naphthalene	(T)
K024	Distillation bottoms from the production of phthalic anhydride from naphthalene	(T)
K093	Distillation light ends from the production of phthalic anhydride from ortho-xylene	(T)
K094	Distillation bottoms from the production of phthalic anhydride from ortho-xylene...........	(T)
K025	Distillation bottoms from the production of nitrobenzene by the nitration of benzene.....	(T)
K026	Stripping still tails from the production of methy ethyl pyridines	(T)
K027	Centrifuge and distillation residues from toluene diisocyanate production....................	(R, T)
K028	Spent catalyst from the hydrochlorinator reactor in the production of 1,1,1-trichloroethane.	(T)
K029	Waste from the product steam stripper in the production of 1,1,1-trichloroethane	(T)
K095	Distillation bottoms from the production of 1,1,1-trichloroethane...............................	(T)
K096	Heavy ends from the heavy ends column from the production of 1,1,1-trichloroethane.	(T)
K030	Column bottoms or heavy ends from the combined production of trichloroethylene and perchloroethylene.	(T)
K083	Distillation bottoms from aniline production	(T)
K103	Process residues from aniline extraction from the production of aniline.......................	(T)
K104	Combined wastewater streams generated from nitrobenzene/aniline production	(T)
K085	Distillation or fractionation column bottoms from the production of chlorobenzenes.......	(T)
K105	Separated aqueous stream from the reactor product washing step in the production of chlorobenzenes.	(T)

Industry and EPA hazardous waste No.	Hazardous waste	Hazard code
Inorganic chemicals:		
K071	Brine purification muds from the mercury cell process in chlorine production, where separately prepurified brine is not used.	(T)
K073	Chlorinated hydrocarbon waste from the purification step of the diaphragm cell process using graphite anodes in chlorine production.	(T)
K106	Wastewater treatment sludge from the mercury cell process in chlorine production......	(T)
Pesticides:		
K031	By-product salts generated in the production of MSMA and cacodylic acid	(T)
K032	Wastewater treatment sludge from the production of chlordane....................................	(T)
K033	Wastewater and scrub water from the chlorination of cyclopentadiene in the production of chlordane.	(T)
K034	Filter solids from the filtration of hexachlorocyclopentadiene in the production of chlordane.	(T)
K097	Vacuum stripper discharge from the chlordane chlorinator in the production of chlordane.	(T)
K035	Wastewater treatment sludges generated in the production of creosote.....................	(T)
K036	Still bottoms from toluene reclamation distillation in the production of disulfoton...........	(T)
K037	Wastewater treatment sludges from the production of disulfoton..............................	(T)
K038	Wastewater from the washing and stripping of phorate production............................	(T)
K039	Filter cake from the filtration of diethylphosphorodithioic acid in the production of phorate.	(T)
K040	Wastewater treatment sludge from the production of phorate	(T)
K041	Wastewater treatment sludge from the production of toxaphene	(T)
K098	Untreated process wastewater from the production of toxaphene	(T)
K042	Heavy ends or distillation residues from the distillation of tetrachlorobenzene in the production of 2,4,5-T.	(T)
K043	2,6-Dichlorophenol waste from the production of 2,4-D	(T)
K099	Untreated wastewater from the production of 2,4-D	(T)
Explosives:		
K044	Wastewater treatment sludges from the manufacturing and processing of explosives ...	(R)
K045	Spent carbon from the treatment of wastewater containing explosives........................	(R)
K046	Wastewater treatment sludges from the manufacturing, formulation and loading of lead-based initiating compounds.	(T)
K047	Pink/red water from TNT operations	(R)
Petroleum refining:		
K048	Dissolved air flotation (DAF) float from the petroleum refining industry........................	(T)
K049	Slop oil emulsion solids from the petroleum refining industry....................................	(T)
K050	Heat exchanger bundle cleaning sludge from the petroleum refining industry................	(T)
K051	API separator sludge from the petroleum refining industry	(T)
K052	Tank bottoms (leaded) from the petroleum refining industry	(T)
Iron and steel:		
K061	Emission control dust/sludge from the primary production of steel in electric furnaces.	(T)
K062	Spent pickle liquor from steel finishing operations	(C, T)
Secondary lead:		
K069	Emission control dust/sludge from secondary lead smelting...................................	(T)
K100	Waste leaching solution from acid leaching of emission control dust/sludge from secondary lead smelting.	(T)
Veterinary pharmaceuticals:		
K084	Wastewater treatment sludges generated during the production of veterinary pharmaceuticals from arsenic or organo-arsenic compounds.	(T)
K101	Distillation tar residues from the distillation of aniline-based compounds in the production of veterinary pharmaceuticals from arsenic or organo-arsenic compounds.	(T)
K102	Residue from the use of activated carbon for decolorization in the production of veterinary pharmaceuticals from arsenic or organo-arsenic compounds.	(T)
Ink formulation: K086	Solvent washes and sludges, caustic washes and sludges, or water washes and sludges from cleaning tubs and equipment used in the formulation of ink from pigments, driers, soaps, and stabilizers containing chromium and lead.	(T)
Coking:		
K060	Ammonia still lime sludge from coking operations..	(T)
K087	Decanter tank tar sludge from coking operations..	(T)

[46 FR 4618, Jan. 16, 1981, as amended at 46 FR 27476–27477, May 20, 1981]

§ 261.33 Discarded commercial chemical products, off-specification species, container residues, and spill residues thereof.

The following materials or items are hazardous wastes if and when they are discarded or intended to be discarded:

(a) Any commercial chemical product, or manufacturing chemical intermediate having the generic name listed in paragraph (e) or (f) of this section.

(b) Any off-specification commercial chemical product or manufacturing chemical intermediate which, if it met specifications, would have the generic name listed in paragraph (e) or (f) of this section.

(c) Any residue remaining in a container or an inner liner removed from a container that has held any commercial chemical product or manufacturing chemical intermediate having the generic name listed in paragraph (e) of this section, unless the container is empty as defined in § 261.7(b)(3) of this chapter.

[*Comment:* Unless the residue is being beneficially used or reused, or legitimately recycled or reclaimed; or being accumulated, stored, transported or treated prior to such use, re-use, recycling or reclamation, EPA considers the residue to be intended for discard, and thus a hazardous waste. An example of a legitimate re-use of the residue would be where the residue remains in the container and the container is used to hold the same commerical chemical product or manufacturing chemical product or manufacturing chemical intermediate it previously held. An example of the discard of the residue would be where the drum is sent to a drum reconditioner who reconditions the drum but discards the residue.]

(d) Any residue or contaminated soil, water or other debris resulting from the cleanup of a spill into or on any land or water of any commercial chemical product or manufacturing chemical intermediate having the generic name listed in paragraph (e) or (f) of this section, or any residue or contaminated soil, water or other debris resulting from the cleanup of a spill, into or on any land or water, of any off-specification chemical product and manufacturing chemical intermediate which, if it met specifications, would have the generic name listed in paragraph (e) or (f) of this section.

[*Comment:* The phrase "commercial chemical product or manufacturing chemical intermediate having the generic name listed in . . ." refers to a chemical substance which is manufactured or formulated for commercial or manufacturing use which consists of the commercially pure grade of the chemical, any technical grades of the chemical that are produced or marketed, and all formulations in which the chemical is the sole active ingredient. It does not refer to a material, such as a manufacturing process waste, that contains any of the substances listed in paragraphs (e) or (f). Where a manufacturing process waste is deemed to be a hazardous waste because it contains a substance listed in paragraphs (e) or (f), such waste will be listed in either §§ 261.31 or 261.32 or will be identified as a hazardous waste by the characteristics set forth in Subpart C of this Part.]

(e) The commercial chemical products, manufacturing chemical intermediates or off-specification commercial chemical products or manufacturing chemical intermediates referred to in paragraphs (a) through (d) of this section, are identified as acute hazardous wastes (H) and are subject to be the small quantity exclusion defined in § 261.5(e).

[*Comment:* For the convenience of the regulated community the primary hazardous properties of these materials have been indicated by the letters T (Toxicity), and R (Reactivity). Absence of a letter indicates that the compound only is listed for acute toxicity.]

These wastes and their corresponding EPA Hazardous Waste Numbers are:

Hazardous waste No.	Substance
P023	Acetaldehyde, chloro-
P002	Acetamide, N-(aminothioxomethyl)-
P057	Acetamide, 2-fluoro-
P058	Acetic acid, fluoro-, sodium salt
P066	Acetimidic acid, N-[(methylcarbamoyl)oxy]thio-, methyl ester
P001	3-(alpha-acetonylbenzyl)-4-hydroxycoumarin and salts
P002	1-Acetyl-2-thiourea
P003	Acrolein
P070	Aldicarb
P004	Aldrin
P005	Allyl alcohol
P006	Aluminum phosphide
P007	5-(Aminomethyl)-3-isoxazolol
P008	4-aAminopyridine
P009	Ammonium picrate (R)
P119	Ammonium vanadate
P010	Arsenic acid
P012	Arsenic (III) oxide
P011	Arsenic (V) oxide
P011	Arsenic pentoxide
P012	Arsenic trioxide
P038	Arsine, diethyl-
P054	Aziridine
P013	Barium cyanide
P024	Benzenamine, 4-chloro-
P077	Benzenamine, 4-nitro-
P028	Benzene, (chloromethyl)-
P042	1,2-Benzenediol, 4-[1-hydroxy-2-(methylamino)ethyl]-

Hazardous waste No.	Substance
P014	Benzenethiol
P028	Benzyl chloride
P015	Beryllium dust
P016	Bis(chloromethyl) ether
P017	Bromoacetone
P018	Brucine
P021	Calcium cyanide
P123	Camphene, octachloro-
P103	Carbamimidoselenoic acid
P022	Carbon bisulfide
P022	Carbon disulfide
P095	Carbonyl chloride
P033	Chlorine cyanide
P023	Chloroacetaldehyde
P024	p-Chloroaniline
P026	1-(o-Chlorophenyl)thiourea
P027	3-Chloropropionitrile
P029	Copper cyanides
P030	Cyanides (soluble cyanide salts), not elsewhere specified
P031	Cyanogen
P033	Cyanogen chloride
P036	Dichlorophenylarsine
P037	Dieldrin
P038	Diethylarsine
P039	O,O-Diethyl S-[2-(ethylthio)ethyl] phosphorodithioate
P041	Diethyl-p-nitrophenyl phosphate
P040	O,O-Diethyl O-pyrazinyl phosphorothioate
P043	Diisopropyl fluorophosphate
P044	Dimethoate
P045	3,3-Dimethyl-1-(methylthio)-2-butanone, O-[(methylamino)carbonyl] oxime
P071	O,O-Dimethyl O-p-nitrophenyl phosphorothioate
P082	Dimethylnitrosamine
P046	alpha, alpha-Dimethylphenethylamine
P047	4,6-Dinitro-o-cresol and salts
P034	4,6-Dinitro-o-cyclohexylphenol
P048	2,4-Dinitrophenol
P020	Dinoseb
P085	Diphosphoramide, octamethyl-
P039	Disulfoton
P049	2,4-Dithiobiuret
P109	Dithiopyrophosphoric acid, tetraethyl ester
P050	Endosulfan
P088	Endothall
P051	Endrin
P042	Epinephrine
P046	Ethanamine, 1,1-dimethyl-2-phenyl-
P084	Ethenamine, N-methyl-N-nitroso-
P101	Ethyl cyanide
P054	Ethylenimine
P097	Famphur
P056	Fluorine
P057	Fluoroacetamide
P058	Fluoroacetic acid, sodium salt
P065	Fulminic acid, mercury(II) salt (R,T)
P059	Heptachlor
P051	1,2,3,4,10,10-Hexachloro-6,7-epoxy-1,4,4a,5,6,7,8,8a-octahydro-endo,endo-1,4:5,8-dimethanonaphthalene
P037	1,2,3,4,10,10-Hexachloro-6,7-epoxy-1,4,4a,5,6,7,8,8a-octahydro-endo,exo-1,4:5,8-dimethanonaphthalene
P060	1,2,3,4,10,10-Hexachloro-1,4,4a,5,8,8a-hexahydro-1,4:5,8-endo, endo-dimeth- anonaphthalene
P004	1,2,3,4,10,10-Hexachloro-1,4,4a,5,8,8a-hexahydro-1,4:5,8-endo,exo-dimethanonaphthalene

Hazardous waste No.	Substance
P060	Hexachlorohexahydro-exo,exo-dimethanonaphthalene
P062	Hexaethyl tetraphosphate
P116	Hydrazinecarbothioamide
P068	Hydrazine, methyl-
P063	Hydrocyanic acid
P063	Hydrogen cyanide
P096	Hydrogen phosphide
P064	Isocyanic acid, methyl ester
P007	3(2H)-Isoxazolone, 5-(aminomethyl)-
P092	Mercury, (acetato-O)phenyl-
P065	Mercury fulminate (R,T)
P016	Methane, oxybis(chloro-
P112	Methane, tetranitro- (R)
P118	Methanethiol, trichloro-
P059	4,7-Methano-1H-indene, 1,4,5,6,7,8,8-heptachloro-3a,4,7,7a-tetrahydro-
P066	Methomyl
P067	2-Methylaziridine
P068	Methyl hydrazine
P064	Methyl isocyanate
P069	2-Methyllactonitrile
P071	Methyl parathion
P072	alpha-Naphthylthiourea
P073	Nickel carbonyl
P074	Nickel cyanide
P074	Nickel(II) cyanide
P073	Nickel tetracarbonyl
P075	Nicotine and salts
P076	Nitric oxide
P077	p-Nitroaniline
P078	Nitrogen dioxide
P076	Nitrogen(II) oxide
P078	Nitrogen(IV) oxide
P081	Nitroglycerine (R)
P082	N-Nitrosodimethylamine
P084	N-Nitrosomethylvinylamine
P050	5-Norbornene-2,3-dimethanol, 1,4,5,6,7,7-hexachloro, cyclic sulfite
P085	Octamethylpyrophosphoramide
P087	Osmium oxide
P087	Osmium tetroxide
P088	7-Oxabicyclo[2.2.1]heptane-2,3-dicarboxylic acid
P089	Parathion
P034	Phenol, 2-cyclohexyl-4,6-dinitro-
P048	Phenol, 2,4-dinitro-
P047	Phenol, 2,4-dinitro-6-methyl-
P020	Phenol, 2,4-dinitro-6-(1-methylpropyl)-
P009	Phenol, 2,4,6-trinitro-, ammonium salt (R)
P036	Phenyl dichloroarsine
P092	Phenylmercuric acetate
P093	N-Phenylthiourea
P094	Phorate
P095	Phosgene
P096	Phosphine
P041	Phosphoric acid, diethyl p-nitrophenyl ester
P044	Phosphorodithioic acid, O,O-dimethyl S-[2-(methylamino)-2-oxoethyl]ester
P043	Phosphorofluoric acid, bis(1-methylethyl)-ester
P094	Phosphorothioic acid, O,O-diethyl S-(ethylthio)methyl ester
P089	Phosphorothioci acid, O,O-diethyl O-(p-nitrophenyl) ester
P040	Phosphorothioic acid, O,O-diethyl O- pyrazinyl ester
P097	Phosphorothioic acid, O,O-dimethyl O-[p-((dimethylamino)-sulfonyl)phenyl]ester
P110	Plumbane, tetraethyl-
P098	Potassium cyanide
P099	Potassium silver cyanide

Hazardous waste No.	Substance
P070	Propanal, 2-methyl-2-(methylthio)-, O-[(methylamino)carbonyl]oxime
P101	Propanenitrile
P027	Propanenitrile, 3-chloro-
P069	Propanenitrile, 2-hydroxy-2-methyl-
P081	1,2,3-Propanetriol, trinitrate- (R)
P017	2-Propanone, 1-bromo-
P102	Propargyl alcohol
P003	2-Propenal
P005	2-Propen-1-ol
P067	1,2-Propylenimine
P102	2-Propyn-1-ol
P008	4-Pyridinamine
P075	Pyridine, (S)-3-(1-methyl-2-pyrrolidinyl)-, and salts
P111	Pyrophosphoric acid, tetraethyl ester
P103	Selenourea
P104	Silver cyanide
P105	Sodium azide
P106	Sodium cyanide
P107	Strontium sulfide
P108	Strychnidin-10-one, and salts
P018	Strychnidin-10-one, 2,3-dimethoxy-
P108	Strychnine and salts
P115	Sulfuric acid, thallium(I) salt
P109	Tetraethyldithiopyrophosphate
P110	Tetraethyl lead
P111	Tetraethylpyrophosphate
P112	Tetranitromethane (R)
P062	Tetraphosphoric acid, hexaethyl ester
P113	Thallic oxide
P113	Thallium(III) oxide
P114	Thallium(I) selenite
P115	Thallium(I) sulfate
P045	Thiofanox
P049	Thioimidodicarbonic diamide
P014	Thiophenol
P116	Thiosemicarbazide
P026	Thiourea, (2-chlorophenyl)-
P072	Thiourea, 1-naphthalenyl-
P093	Thiourea, phenyl-
P123	Toxaphene
P118	Trichloromethanethiol
P119	Vanadic acid, ammonium salt
P120	Vanadium pentoxide
P120	Vanadium(V) oxide
P001	Warfarin
P121	Zinc cyanide
P122	Zinc phosphide (R,T)

(f) The commercial chemical products, manufacturing chemical intermediates, or off-specification commercial chemical products referred to in paragraphs (a) through (d) of this section, are identified as toxic wastes (T) unless otherwise designated and are subject to the small quantity exclusion defined in § 261.5 (a) and (f).

[*Comment:* For the convenience of the regulated community, the primary hazardous properties of these materials have been indicated by the letters T (Toxicity), R (Reactivity), I (Ignitability) and C (Corrosivity). Absence of a letter indicates that the compound is only listed for toxicity.]

§ 261.33

These wastes and their corresponding EPA Hazardous Waste Numbers are:

Hazardous Waste No.	Substance
U001	Acetaldehyde (I)
U034	Acetaldehyde, trichloro-
U187	Acetamide, N-(4-ethoxyphenyl)-
U005	Acetamide, N-9H-fluoren-2-yl
U112	Acetic acid, ethyl ester (I)
U144	Acetic acid, lead salt
U214	Acetic acid, thallium(I) salt
U002	Acetone (I)
U003	Acetonitrile (I,T)
U004	Acetophenone
U005	2-Acetylaminofluorene
U006	Acetyl chloride (C,R,T)
U007	Acrylamide
U008	Acrylic acid (I)
U009	Acrylonitrile
U150	Alanine, 3-[p-bis(2-chloroethyl)amino] phenyl-, L-
U011	Amitrole
U012	Aniline (I,T)
U014	Auramine
U015	Azaserine
U010	Azirino(2',3':3,4)pyrrolo(1,2-a)indole-4,7-dione, 6-amino-8-[((aminocarbonyl) oxy)methyl]-1,1a,2,8,8a,8b-hexahydro-8a-methoxy-5-methyl-,
U157	Benz[j]aceanthrylene, 1,2-dihydro-3-methyl-
U016	Benz[c]acridine
U016	3,4-Benzacridine
U017	Benzal chloride
U018	Benz[a]anthracene
U018	1,2-Benzanthracene
U094	1,2-Benzanthracene, 7,12-dimethyl-
U012	Benzenamine (I,T)
U014	Benzenamine, 4,4'-carbonimidoylbis(N,N-di-methyl-
U049	Benzenamine, 4-chloro-2-methyl-
U093	Benzenamine, N,N'-dimethyl-4-phenylazo-
U158	Benzenamine, 4,4'-methylenebis(2-chloro-
U222	Benzenamine, 2-methyl-, hydrochloride
U181	Benzenamine, 2-methyl-5-nitro
U019	Benzene (I,T)
U038	Benzeneacetic acid, 4-chloro-alpha-(4-chloro-phenyl)-alpha-hydroxy, ethyl ester
U030	Benzene, 1-bromo-4-phenoxy-
U037	Benzene, chloro-
U190	1,2-Benzenedicarboxylic acid anhydride
U028	1,2-Benzenedicarboxylic acid, [bis(2-ethyl-hexyl)] ester
U069	1,2-Benzenedicarboxylic acid, dibutyl ester
U088	1,2-Benzenedicarboxylic acid, diethyl ester
U102	1,2-Benzenedicarboxylic acid, dimethyl ester
U107	1,2-Benzenedicarboxylic acid, di-n-octyl ester
U070	Benzene, 1,2-dichloro-
U071	Benzene, 1,3-dichloro-
U072	Benzene, 1,4-dichloro-
U017	Benzene, (dichloromethyl)-
U223	Benzene, 1,3-diisocyanatomethyl- (R,T)
U239	Benzene, dimethyl-(I,T)
U201	1,3-Benzenediol
U127	Benzene, hexachloro-
U056	Benzene, hexahydro- (I)
U188	Benzene, hydroxy-
U220	Benzene, methyl-
U105	Benzene, 1-methyl-1-2,4-dinitro-
U106	Benzene, 1-methyl-2,6-dinitro-
U203	Benzene, 1,2-methylenedioxy-4-allyl-
U141	Benzene, 1,2-methylenedioxy-4-propenyl-

Hazardous Waste No.	Substance
U090	Benzene, 1,2-methylenedioxy-4-propyl-
U055	Benzene, (1-methylethyl)- (I)
U169	Benzene, nitro- (I,T)
U183	Benzene, pentachloro-
U185	Benzene, pentachloro-nitro-
U020	Benzenesulfonic acid chloride (C,R)
U020	Benzenesulfonyl chloride (C,R)
U207	Benzene, 1,2,4,5-tetrachloro-
U023	Benzene, (trichloromethyl)-(C,R,T)
0234	Benzene, 1,3,5-trinitro- (R,T)
U021	Benzidine
U202	1,2-Benzisothiazolin-3-one, 1,1-dioxide
U120	Benzo[j,k]fluorene
U022	Benzo[a]pyrene
U022	3,4-Benzopyrene
U197	p-Benzoquinone
U023	Benzotrichloride (C,R,T)
U050	1,2-Benzphenanthrene
U085	2,2'-Bioxirane (I,T)
U021	(1,1'-Biphenyl)-4,4'-diamine
U073	(1,1'-Biphenyl)-4,4'-diamine, 3,3'-dichloro-
U091	(1,1'-Biphenyl)-4,4'-diamine, 3,3'-dimethoxy-
U095	(1,1'-Biphenyl)-4,4'-diamine, 3,3'-dimethyl-
U024	Bis(2-chloroethoxy) methane
U027	Bis(2-chloroisopropyl) ether
U244	Bis(dimethylthiocarbamoyl) disulfide
U028	Bis(2-ethylhexyl) phthalate
U246	Bromine cyanide
U225	Bromoform
U030	4-Bromophenyl phenyl ether
U128	1,3-Butadiene, 1,1,2,3,4,4-hexachloro-
U172	1-Butanamine, N-butyl-N-nitroso-
U035	Butanoic acid, 4-[Bis(2-chloroethyl)amino] benzene-
U031	1-Butanol (I)
U159	2-Butanone (I,T)
U160	2-Butanone peroxide (R,T)
U053	2-Butenal
U074	2-Butene, 1,4-dichloro- (I,T)
U031	n-Butyl alchohol (I)
U136	Cacodylic acid
U032	Calcium chromate
U238	Carbamic acid, ethyl ester
U178	Carbamic acid, methylnitroso-, ethyl ester
U176	Carbamide, N-ethyl-N-nitroso-
U177	Carbamide, N-methyl-N-nitroso-
U219	Carbamide, thio-
U097	Carbamoyl chloride, dimethyl-
U215	Carbonic acid, dithallium(I) salt
U156	Carbonochloridic acid, methyl ester (I,T)
U033	Carbon oxyfluoride (R,T)
U211	Carbon tetrachloride
U033	Carbonyl fluoride (R,T)
U034	Chloral
U035	Chlorambucil
U036	Chlordane, technical
U026	Chlornaphazine
U037	Chlorobenzene
U039	4-Chloro-m-cresol
U041	1-Chloro-2,3-epoxypropane
U042	2-Chloroethyl vinyl ether
U044	Chloroform
U046	Chloromethyl methyl ether
U047	beta-Chloronaphthalene
U048	o-Chlorophenol
U049	4-Chloro-o-toluidine, hydrochloride
U032	Chromic acid, calcium salt
U050	Chrysene
U051	Creosote
U052	Cresols
U052	Cresylic acid
U053	Crotonaldehyde

Hazardous Waste No.	Substance
U055	Cumene (I)
U246	Cyanogen bromide
U197	1,4-Cyclohexadienedione
U056	Cyclohexane (I)
U057	Cyclohexanone (I)
U130	1,3-Cyclopentadiene, 1,2,3,4,5,5-hexa- chloro-
U058	Cyclophosphamide
U240	2,44-D, salts and esters
U059	Daunomycin
U060	DDD
U061	DDT
U142	Decachlorooctahydro-1,3,4-metheno-2H-cyclobuta[c,d]-pentalen-2-one
U062	Diallate
U133	Diamine (R,T)
U221	Diaminotoluene
U063	Dibenz[a,h]anthracene
U063	1,2:5,6-Dibenzanthracene
U064	1,2:7,8-Dibenzopyrene
U064	Dibenz[a,i]pyrene
U066	1,2-Dibromo-3-chloropropane
U069	Dibutyl phthalate
U062	S-(2,3-Dichloroallyl) diisopropylthiocarbamate
U070	o-Dichlorobenzene
U071	m-Dichlorobenzene
U072	p-Dichlorobenzene
U073	3,3'-Dichlorobenzidine
U074	1,4-Dichloro-2-butene (I,T)
U075	Dichlorodifluoromethane
U192	3,5-Dichloro-N-(1,1-dimethyl-2-propynyl) benzamide
U060	Dichloro diphenyl dichloroethane
U061	Dichloro diphenyl trichloroethane
U078	1,1-Dichloroethylene
U079	1,2-Dichloroethylene
U025	Dichloroethyl ether
U081	2,4-Dichlorophenol
U082	2,6-Dichlorophenol
U240	2,4-Dichlorophenoxyacetic acid, salts and esters
U083	1,2-Dichloropropane
U084	1,3-Dichloropropene
U085	1,2:3,4-Diepoxybutane (I,T)
U108	1,4-Diethylene dioxide
U086	N,N-Diethylhydrazine
U087	O,O-Diethyl-S-methyl-dithiophosphate
U088	Diethyl phthalate
U089	Diethylstilbestrol
U148	1,2-Dihydro-3,6-pyradizinedione
U090	Dihydrosafrole
U091	3,3'-Dimethoxybenzidine
U092	Dimethylamine (I)
U093	Dimethylaminoazobenzene
U094	7,12-Dimethylbenz[a]anthracene
U095	3,3'-Dimethylbenzidine
U096	alpha,alpha-Dimethylbenzylhydroperoxide (R)
U097	Dimethylcarbamoyl chloride
U098	1,1-Dimethylhydrazine
U099	1,2-Dimethylhydrazine
U101	2,4-Dimethylphenol
U102	Dimethyl phthalate
U103	Dimethyl sulfate
U105	2,4-Dinitrotoluene
U106	2,6-Dinitrotoluene
U107	Di-n-octyl phthalate
U108	1,4-Dioxane
U109	1,2- Diphenylhydrazine
U110	Dipropylamine (I)
U111	Di-N-propylnitrosamine
U001	Ethanal (I)
U174	Ethanamine, N-ethyl-N-nitroso-
U067	Ethane, 1,2-dibromo-

Hazardous Waste No.	Substance
U076	Ethane, 1,1-dichloro-
U077	Ethane, 1,2-dichloro-
U114	1,2-Ethanediylbiscarbamodithioic acid
U131	Ethane, 1,1,1,2,2,2-hexachloro-
U024	Ethane, 1,1'-[methylenebis(oxy)]bis[2-chloro-
U003	Ethanenitrile (I, T)
U117	Ethane,1,1'-oxybis- (I)
U025	Ethane, 1,1'-oxybis[2-chloro-
U184	Ethane, pentachloro-
U208	Ethane, 1,1,1,2-tetrachloro-
U209	Ethane, 1,1,2,2-tetrachloro-
U218	Ethanethioamide
U247	Ethane, 1,1,1,-trichloro-2,2-bis(p-methoxyphenyl).
U227	Ethane, 1,1,2-trichloro-
U043	Ethene, chloro-
U042	Ethene, 2-chloroethoxy-
U078	Ethene, 1,1-dichloro-
U079	Ethene, trans-1,2-dichloro-
U210	Ethene, 1,1,2,2-tetrachloro-
U173	Ethanol, 2,2'-(nitrosoimino)bis-
U004	Ethanone, 1-phenyl-
U006	Ethanoyl chloride (C,R,T)
U112	Ethyl acetate (I)
U113	Ethyl acrylate (I)
U238	Ethyl carbamate (urethan)
U038	Ethyl 4,4'-dichlorobenzilate
U114	Ethylenebis(dithiocarbamic acid)
U067	Etylene dibromide
U077	Ethylene dichloride
U115	Ethlene oxide (I,T)
U116	Ethylene thiourea
U117	Ethyl ether (I)
U076	Ethylidene dichloride
U118	Ethylmethacrylate
U119	Ethyl methanesulfonate
U139	Ferric dextran
U120	Fluoranthene
U122	Formaldehyde
U123	Formic acid (C,T)
U124	Furan (I)
U125	2-Furancarboxaldehyde (I)
U147	2,5-Furandione
U213	Furan, tetrahydro- (I)
U125	Furfural (I)
U124	Furfuran (I)
U206	D-Glucopyranose, 2-deoxy-2(3-methyl-3-nitrosoureido)-
U126	Glycidylaldehyde
U163	Guanidine, N-nitroso-N-methyl-N'nitro-
U127	Hexachlorobenzene
U128	Hexachlorobutadiene
U129	Hexachlorocyclohexane (gamma isomer)
U130	Hexachlorocyclopentadiene
U131	Hexachloroethane
U132	Hexachlorophene
U243	Hexachloropropene
U133	Hydrazine (R,T)
U086	Hydrazine, 1,2-diethyl-
U098	Hydrazine, 1,1-dimethyl-
U099	Hydrazine, 1,2-dimethyl-
U109	Hydrazine, 1,2-diphenyl-
U134	Hydrofluoric acid (C,T)
U134	Hydrogen fluoride (C,T)
U135	Hydrogen sulfide
U096	Hydroperoxide, 1-methyl-1-phenylethyl- (R)
U136	Hydroxydimethylarsine oxide
U116	2-Imidazolidinethione
U137	Indeno[1,2,3-cd]pyrene
U139	Iron dextran
U140	Isobutyl alcohol (I,T)
U141	Isosafrole

Hazardous Waste No.	Substance
U142	Kepone
U143	Lasiocarpine
U144	Lead acetate
U145	Lead phosphate
U146	Lead subacetate
U129	Lindane
U147	Maleic anhydride
U148	Maleic hydrazide
U149	Malononitrile
U150	Melphalan
U151	Mercury
U152	Methacrylonitrile (I,T)
U092	Methanamine, N-methyl- (I)
U029	Methane, bromo-
U045	Methane, chloro- (I,T)
U046	Methane, chloromethoxy-
U068	Methane, dibromo-
U080	Methane, dichloro-
U075	Methane, dichlorodifluoro-
U138	Methane, iodo-
U119	Methanesulfonic acid, ethyl ester
U211	Methane, tetrachloro-
U121	Methane, trichlorofluoro-
U153	Methanethiol (I,T)
U225	Methane, tribromo-
U044	Methane, trichloro-
U121	Methane, trichlorofluoro-
U123	Methanoic acid (C,T)
U036	4,7-Methanoindan, 1,2,4,5,6,7,8,8-octa-chloro-3a,4,7,7a-tetrahydro-
U154	Methanol (I)
U155	Methapyrilene
U247	Methoxychlor.
U154	Methyl alcohol (I)
U029	Methyl bromide
U186	1-Methylbutadiene (I)
U045	Methyl chloride (I,T)
U156	Methyl chlorocarbonate (I,T)
U226	Methylchloroform
U157	3-Methylcholanthrene
U158	4,4'-Methylenebis(2-chloroaniline)
U132	2,2'-Methylenebis(3,4,6-trichlorophenol)
U068	Methylene bromide
U080	Methylene chloride
U122	Methylene oxide
U159	Methyl ethyl ketone (I,T)
U160	Methyl ethyl ketone peroxide (R,T)
U138	Methyl iodide
U161	Methyl isobutyl ketone (I)
U162	Methyl methacrylate (I,T)
U163	N-Methyl-N'-nitro-N-nitrosoguanidine
U161	4-Methyl-2-pentanone (I)
U164	Methylthiouracil
U010	Mitomycin C
U059	5,12-Naphthacenedione, (8S-cis)-8-acetyl-10-[(3-amino-2,3,6-trideoxy-alpha-L-lyxo-hexopyranosyl)oxy]-7,8,9,10-tetrahydro-6,8,11-trihydroxy-1-methoxy-
U165	Naphthalene
U047	Naphthalene, 2-chloro-
U166	1,4-Naphthalenedione
U236	2,7-Naphthalenedisulfonic acid, 3,3'-[(3,3'-di-methyl-(1,1'-biphenyl)-4,4'diyl)]-bis(azo)bis(5-amino-4-hydroxy)-,tetrasodium salt
U166	1,4,Naphthaquinone
U167	1-Naphthylamine
U168	2-Naphthylamine
U167	alpha-Naphthylamine
U168	beta-Naphthylamine
U026	2-Naphthylamine, N,N'-bis(2-chloromethyl)-
U169	Nitrobenzene (I,T)

Hazardous Waste No.	Substance
U170	p-Nitrophenol
U171	2-Nitropropane (I)
U172	N-Nitrosodi-n-butylamine
U173	N-Nitrosodiethanolamine
U174	N-Nitrosodiethylamine
U111	N-Nitroso-N-propylamine
U176	N-Nitroso-N-ethylurea
U177	N-Nitroso-N-methylurea
U178	N-Nitroso-N-methylurethane
U179	N-Nitrosopiperidine
U180	N-Nitrosopyrrolidine
U181	5-Nitro-o-toluidine
U193	1,2-Oxathiolane, 2,2-dioxide
U058	2H-1,3,2-Oxazaphosphorine, 2-[bis(2-chloro-ethyl)amino]tetrahydro-, oxide 2-
U115	Oxirane (I,T)
U041	Oxirane, 2-(chloromethyl)-
U182	Paraldehyde
U183	Pentachlorobenzene
U184	Pentachloroethane
U185	Pentachloronitrobenzene
U242	Pentachlorophenol
U186	1,3-Pentadiene (I)
U187	Phenacetin
U188	Phenol
U048	Phenol, 2-chloro-
U039	Phenol, 4-chloro-3-methyl-
U081	Phenol, 2,4-dichloro-
U082	Phenol, 2,6-dichloro-
U101	Phenol, 2,4-dimethyl-
U170	Phenol, 4-nitro-
U242	Phenol, pentachloro-
U212	Phenol, 2,3,4,6-tetrachloro-
U230	Phenol, 2,4,5-trichloro-
U231	Phenol, 2,4,6-trichloro-
U137	1,10-(1,2-phenylene)pyrene
U145	Phosphoric acid, Lead salt
U087	Phosphorodithioic acid, O,O-diethyl-, S-methy-lester
U189	Phosphorous sulfide (R)
U190	Phthalic anhydride
U191	2-Picoline
U192	Pronamide
U194	1-Propanamine (I,T)
U110	1-Propanamine, N-propyl- (I)
U066	Propane, 1,2-dibromo-3-chloro-
U149	Propanedinitrile
U171	Propane, 2-nitro- (I)
U027	Propane, 2,2'oxybis[2-chloro-
U193	1,3-Propane sultone
U235	1-Propanol, 2,3-dibromo-, phosphate (3:1)
U126	1-Propanol, 2,3-epoxy-
U140	1-Propanol, 2-methyl- (I,T)
U002	2-Propanone (I)
U007	2-Propenamide
U084	Propene, 1,3-dichloro-
U243	1-Propene, 1,1,2,3,3,3-hexachloro-
U009	2-Propenenitrile
U152	2-Propenenitrile, 2-methyl- (I,T)
U008	2-Propenoic acid (I)
U113	2-Propenoic acid, ethyl ester (I)
U118	2-Propenoic acid, 2-methyl-, ethyl ester
U162	2-Propenoic acid, 2-methyl-, methyl ester (I,T)
U233	Propionic acid, 2-(2,4,5-trichlorophenoxy)-
U194	n-Propylamine (I,T)
U083	Propylene dichloride
U196	Pyridine
U155	Pyridine, 2-[(2-(dimethylamino)-2-thenyla-mino]-
U179	Pyridine, hexahydro-N-nitroso-
U191	Pyridine, 2-methyl-

Hazardous Waste No.	Substance
U164	4(1H)-Pyrimidinone, 2,3-dihydro-6-methyl-2-thioxo-
U180	Pyrrole, tetrahydro-N-nitroso-
U200	Reserpine
U201	Resorcinol
U202	Saccharin and salts
U203	Safrole
U204	Selenious acid
U204	Selenium dioxide
U205	Selenium disulfide (R,T)
U015	L-Serine, diazoacetate (ester)
U233	Silvex
U089	4,4'-Stilbenediol, alpha,alpha'-diethyl-
U206	Streptozotocin
U135	Sulfur hydride
U103	Sulfuric acid, dimethyl ester
U189	Sulfur phosphide (R)
U205	Sulfur selenide (R,T)
U232	2,4,5-T
U207	1,2,4,5-Tetrachlorobenzene
U208	1,1,1,2-Tetrachloroethane
U209	1,1,2,2-Tetrachloroethane
U210	Tetrachloroethylene
U212	2,3,4,6-Tetrachlorophenol
U213	Tetrahydrofuran (I)
U214	Thallium(I) acetate
U215	Thallium(I) carbonate
U216	Thallium(I) chloride
U217	Thallium(I) nitrate
U218	Thioacetamide
U153	Thiomethanol (I,T)
U219	Thiourea
U244	Thiram
U220	Toluene
U221	Toluenediamine
U223	Toluene diisocyanate (R,T)
U222	O-Toluidine hydrochloride
U011	1H-1,2,4-Triazol-3-amine
U226	1,1,1-Trichloroethane
U227	1,1,2-Trichloroethane
U228	Trichloroethene
U228	Trichloroethylene
U121	Trichloromonofluoromethane
U230	2,4,5-Trichlorophenol
U231	2,4,6-Trichlorophenol
U232	2,4,5-Trichlorophenoxyacetic acid
U234	sym-Trinitrobenzene (R,T)
U182	1,3,5-Trioxane, 2,4,5-trimethyl-
U235	Tris(2,3-dibromopropyl) phosphate
U236	Trypan blue
U237	Uracil, 5[bis(2-chloroethyl)amino]-
U237	Uracil mustard
U043	Vinyl chloride
U239	Xylene (I)
U200	Yohimban-16-carboxylic acid, 11,17-dimethoxy-18-[(3,4,5-trimethoxy-benzoyl)oxy]-, methyl ester.

[45 FR 78529, 78541, Nov. 25, 1980, as amended at 46 FR 27477, May 20, 1981]

APPENDIX I—REPRESENTATIVE SAMPLING METHODS

The methods and equipment used for sampling waste materials will vary with the form and consistency of the waste materials to be sampled. Samples collected using the sampling protocols listed below, for sampling waste with properties similar to the in-dicated materials, will be considered by the Agency to be representative of the waste.

Extremely viscous liquid—ASTM Standard D140-70 Crushed or powdered material—ASTM Standard D346-75 Soil or rock-like material—ASTM Standard D420-69 Soil-like material—ASTM Standard D1452-65

Fly Ash-like material—ASTM Standard D2234-76 [ASTM Standards are available from ASTM, 1916 Race St., Philadelphia, PA 19103]

Containerized liquid wastes—"COLIWASA" described in "Test Methods for the Evaluation of Solid Waste, Physical/Chemical Methods," [1] U.S. Environmental Protection Agency, Office of Solid Waste, Washington, D.C. 20460. [Copies may be obtained from Solid Waste Information, U.S. Environmental Protection Agency, 26 W. St. Clair St., Cincinnati, Ohio 45268]

Liquid waste in pits, ponds, lagoons, and similar reservoirs.—"Pond Sampler" described in "Test Methods for the Evaluation of Solid Waste, Physical/Chemical Methods." [1]

This manual also contains additional information on application of these protocols.

APPENDIX II—EP TOXICITY TEST PROCEDURES

A. Extraction Procedure (EP)

1. A representative sample of the waste to be tested (minimum size 100 grams) shall be obtained using the methods specified in Appendix I or any other method capable of yielding a representative sample within the meaning of Part 260. [For detailed guidance on conducting the various aspects of the EP see "Test Methods for the Evaluation of Solid Waste, Physical/Chemical Methods" (incorporated by reference, see § 260.11).]

2. The sample shall be separated into its component liquid and solid phases using the method described in "Separation Procedure" below. If the solid residue [1] obtained using this method totals less than 0.5% of the original weight of the waste, the residue can be discarded and the operator shall treat the liquid phase as the extract and proceed immediately to Step 8.

[1] These methods are also described in "Samplers and Sampling Procedures for Hazardous Waste Streams," EPA 600/2-80-018, January 1980.
[1] The percent solids is determined by drying the filter pad at 80°C until it reaches constant weight and then calculating the percent solids using the following equation:

Percent solids =

$$\frac{(\text{weight of pad} + \text{solid}) - (\text{tare weight of pad})}{\text{initial weight of sample}} \times 100$$

3. The solid material obtained from the Separation Procedure shall be evaluated for its particle size. If the solid material has a surface area per gram of material equal to, or greater than, 3.1 cm² or passes through a 9.5 mm (0.375 inch) standard sieve, the operator shall proceed to Step 4. If the surface area is smaller or the particle size larger than specified above, the solid material shall be prepared for extraction by crushing, cutting or grinding the material so that it passes through a 9.5 mm (0.375 inch) sieve or, if the material is in a single piece, by subjecting the material to the "Structural Integrity Procedure" described below.

4. The solid material obtained in Step 3 shall be weighed and placed in an extractor with 16 times its weight of deionized water. Do not allow the material to dry prior to weighing. For purposes of this test, an acceptable extractor is one which will impart sufficient agitation to the mixture to not only prevent stratification of the sample and extraction fluid but also insure that all sample surfaces are continuously brought into contact with well mixed extraction fluid.

5. After the solid material and deionized water are placed in the extractor, the operator shall begin agitation and measure the pH of the solution in the extractor. If the pH is greater than 5.0, the pH of the solution shall be decreased to 5.0 ± 0.2 by adding 0.5 N acetic acid. If the pH is equal to or less than 5.0, no acetic acid should be added. The pH of the solution shall be monitored, as described below, during the course of the extraction and if the pH rises above 5.2, 0.5N acetic acid shall be added to bring the pH down to 5.0 ± 0.2. However, in no event shall the aggregrate amount of acid added to the solution exceed 4 ml of acid per gram of solid. The mixture shall be agitated for 24 hours and maintained at 20°-40°C (68°-104°F) during this time. It is recommended that the operator monitor and adjust the pH during the course of the extraction with a device such as the Type 45-A pH Controller manufactured by Chemtrix, Inc., Hillsboro, Oregon 97123 or its equivalent, in conjunction with a metering pump and reservoir of 0.5N acetic acid. If such a system is not available, the following manual procedure shall be employed:

(a) A pH meter shall be calibrated in accordance with the manufacturer's specifications.

(b) The pH of the solution shall be checked and, if necessary, 0.5N acetic acid shall be manually added to the extractor until the pH reaches 5.0 ± 0.2. The pH of the solution shall be adjusted at 15, 30 and 60 minute intervals, moving to the next longer interval if the pH does not have to be adjusted more than 0.5N pH units.

(c) The adjustment procedure shall be continued for at least 6 hours.

(d) If at the end of the 24-hour extraction period, the pH of the solution is not below 5.2 and the maximum amount of acid (4 ml per gram of solids) has not been added, the pH shall be adjusted to 5.0 ± 0.2 and the extraction continued for an additional four hours, during which the pH shall be adjusted at one hour intervals.

6. At the end of the 24 hour extraction period, deionized water shall be added to the extractor in an amount determined by the following equation:

$$V = (20)(W) - 16(W) - A$$

V = ml deionized water to be added

W = weight in grams of solid charged to extractor

A = ml of 0.5N acetic acid added during extraction

7. The material in the extractor shall be separated into its component liquid and solid phases as described under "Separation Procedure."

8. The liquids resulting from Steps 2 and 7 shall be combined. This combined liquid (or the waste itself if it has less than ½ percent solids, as noted in step 2) is the extract and shall be analyzed for the presence of any of the contaminants specified in Table I of § 261.24 using the Analytical Procedures designated below.

Separation Procedure

Equipment: A filter holder, designed for filtration media having a nominal pore size of 0.45 micrometers and capable of applying a 5.3 kg/cm² (75 psi) hydrostatic pressure to the solution being filtered, shall be used. For mixtures containing nonabsorptive solids, where separation can be effected without imposing a 5.3 kg/cm² pressure differential, vacuum filters employing a 0.45 micrometers filter media can be used. (For further guidance on filtration equipment or procedures see "Test Methods for Evaluating Solid Waste, Physical/Chemical Methods" incorporated by reference, see § 260.11). Procedure:[2]

[2] This procedure is intended to result in separation of the "free" liquid portion of the waste from any solid matter having a particle size > 0.45 μm. If the sample will not filter, various other separation techniques can be used to aid in the filtration. As described above, pressure filtration is employed to speed up the filtration process. This does not alter the nature of the separation. If liquid does not separate during filtration, the waste can be centrifuged. If sep-

Continued

(i) Following manufacturer's directions, the filter unit shall be assembled with a filter bed consisting of a 0.45 micrometer filter membrane. For difficult or slow to filter mixtures a prefilter bed consisting of the following prefilters in increasing pore size (0.65 micrometer membrane, fine glass fiber prefilter, and coarse glass fiber prefilter) can be used.

(ii) The waste shall be poured into the filtration unit.

(iii) The reservoir shall be slowly pressurized until liquid begins to flow from the filtrate outlet at which point the pressure in the filter shall be immediately lowered to 10-15 psig. Filtration shall be continued until liquid flow ceases.

(iv) The pressure shall be increased stepwise in 10 psi increments to 75 psig and filtration continued until flow ceases or the pressurizing gas begins to exit from the filtrate outlet.

(v) The filter unit shall be depressurized, the solid material removed and weighed and then transferred to the extraction apparatus, or, in the case of final filtration prior to analysis, discarded. Do not allow the material retained on the filter pad to dry prior to weighing.

(vi) The liquid phase shall be stored at 4°C for subsequent use in Step 8.

B. Structural Integrity Procedure

Equipment: A Structural Integrity Tester having a 3.18 cm (1.25 in.) diameter hammer weighing 0.33 kg (0.73 lbs.) and having a

---

aration occurs during centrifugation, the liquid portion (centrifugate) is filtered through the 0.45 μm filter prior to becoming mixed with the liquid portion of the waste obtained from the initial filtration. Any material that will not pass through the filter after centrifugation is considered a solid and is extracted.

free fall of 15.24 cm (6 in.) shall be used. This device is available from Associated Design and Manufacturing Company, Alexandria, VA 22314, as Part No. 125, or it may be fabricated to meet the specifications shown in Figure 1.

Procedure

1. The sample holder shall be filled with the material to be tested. If the sample of waste is a large monolithic block, a portion shall be cut from the block having the dimensions of a 3.3 cm (1.3 in.) diameter x 7.1 cm (2.8 in.) cylinder. For a fixated waste, samples may be cast in the form of a 3.3 cm (1.3 in.) diameter x 7.1 cm (2.8 in.) cylinder for purposes of conducting this test. In such cases, the waste may be allowed to cure for 30 days prior to further testing.

2. The sample holder shall be placed into the Structural Integrity Tester, then the hammer shall be raised to its maximum height and dropped. This shall be repeated fifteen times.

3. The material shall be removed from the sample holder, weighed, and transferred to the extraction apparatus for extraction.

Analytical Procedures for Analyzing Extract Contaminants

The test methods for analyzing the extract are as follows:

1. For arsenic, barium, cadmium, chromium, lead, mercury, selenium, silver, endrin, lindane, methoxychlor, toxaphene, 2,4-D[2,4-dichlorophenoxyacetic acid] or 2,4,5-TP [2,4,5-trichlorophenoxypropionic acid]: "Test Methods for the Evaluation of Solid Waste, Physical/Chemical Methods" (incorporated by reference, see § 260.11).

2. [Reserved]

For all analyses, the methods of standard addition shall be used for quantification of species concentration.

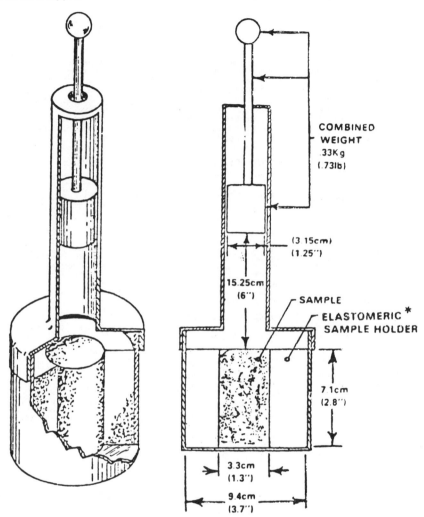

*ELASTOMERIC SAMPLE HOLDER FABRICATED OF
MATERIAL FIRM ENOUGH TO SUPPORT THE SAMPLE

Figure 1
COMPACTION TESTER

[45 FR 33119, May 19, 1980, as amended at 46 FR 35247, July 7, 1981]

APPENDIX III—CHEMICAL ANALYSIS
TEST METHODS

Tables 1, 2, and 3 specify the appropriate analytical procedures described in "Test Methods for Evaluating Solid Waste, Physical/Chemical Methods" (incorporated by reference, see § 260.11), which shall be used in determining whether the waste in question contains a given toxic constituent. Table 1 identifies the analytical class and the approved measurement techniques for each organic chemical listed in Appendix VII. Table 2 identifies the corresponding methods for the inorganic species. Table 3 identifies the specific sample preparation and measurement instrument introduction techniques which may be suitable for both the organic and inorganic species as well as the matrices of concern.

Prior to final selection of the analytical method the operator should consult the specific method descriptions in SW–846 for additional guidance on which of the approved methods should be employed for a specific waste analysis situation.

TABLE 1.—ANALYTICAL CHARACTERISTICS OF ORGANIC CHEMICALS

Compound	Sample handling class/ fraction	Non-GC methods	Measurement techniques		
			GC/MS	Conventional	
				GC	Detector
Acetonitrile	Volatile		8.24	8.03	NSD
Acrolein	Volatile		8.24	8.03	NSD
Acrylamide	Volatile		8.24	8.01	FID
Acrylonitrile	Volatile		8.24	8.03	NSD
Benzene	Volatile		8.24	8.02	PID
Benz(a)anthracene	Extractable/BN	8.10 (HPLC)	8.25	8.10	FID
Benzo(a)pyrene	Extractable/BN	8.10 (HPLC)	8.25	8.10	FID
Benzotrichloride	Extractable/BN		8.25	8.12	ECD
Benzyl chloride	Volatile or Extractable/BN		8.24	8.01	HSD
			8.25	8.12	ECD
Benz(b)fluoanthene	Extractable/BN	8.10 (HPLC)	8.25	8.10	FID
Bis(2-chloroethoxymethane)	Volatile		8.24	8.01	HSD
Bis(2-chloroethyl)ether	Volatile		8.24	8.01	HSD
Bis(2-chloroisopropyl)ether	Volatile		8.24	8.01	HSD
Carbon disulfide	Volatile		8.24	8.01	HSD
Carbon tetrachloride	Volatile		8.24	8.01	HSD
Chlordane	Extractable/BN		8.25	8.08	HSD
Chlorinated dibenzodioxins	Extractable/BN		8.25	8.08	ECD
Chlorinated biphenyls	Extractable/BN		8.25	8.08	HSD
Chloroacetaldehyde	Volatile		8.24	8.01	HSD
Chlorobenzene	Volatile		8.24	8.01	HSD
				8.02	PID
Chloroform	Volatile		8.24	8.01	HSD
Chloromethane	Volatile		8.24	8.01	HSD
2-Chlorophenol	Extractable/BN		8.25	8.04	FID, ECD
Chrysene	Extractable/BN	8.10 (HPLC)	8.25	8.10	FID
Creosote	Extractable/BN		¹8.25	8.10	ECD
Cresol(s)	Extractable/A		8.25	8.04	FID, ECD
Cresylic acid(s)	Extractable/A		8.25	8.04	FID, ECD
Dichlorobenzene(s)	Extractable/BN		8.25	8.01	HSD
				8.02	PID
				8.12	ECD
Dichloroethane(s)	Volatile		8.24	8.01	HSD
Dichloromethane	Volatile		8.24	8.01	HSD
Dichlorophenoxy-acetic acid	Extractable/A		8.25	8.40	HSD
Dichloropropanol	Extractable/BN		8.25	8.12	ECD
2,4-Dimethylphenol	Extractable/A		8.25	8.04	FID, ECD
Dinitrobenzene	Extractable/BN		8.25	8.09	FID, ECD
4,6-Dinitro-o-cresol	Extractable/A		8.25	8.04	FID, ECD
2,4-Dinitrotoluene	Extractable/BN		8.25	8.09	FID, ECD
Endrin	Extractable/P		8.25	8.08	HSD
Ethyl ether	Volatile		8.24	8.01	FID
				8.02	FID
Formaldehyde	Volatile		8.24	8.01	FID
Formic acid	Extractable/BN		8.25	8.06	FID
Heptachlor	Extractable/P		8.25	8.06	HSD
Hexachlorobenzene	Extractable/BN		8.25	8.12	ECD
Hexachlorobutadiene	Extractable/BN		8.25	8.12	ECD
Hexachloroethane	Extractable/BN		8.25	8.12	ECD
Hexachlorocyclopentadiene	Extractable/BN		8.25	8.12	ECD

TABLE 1.—ANALYTICAL CHARACTERISTICS OF ORGANIC CHEMICALS—Continued

Compound	Sample handling class/ fraction	Non-GC methods	Measurement techniques		
			GC/MS	Conventional	
				GC	Detector
Lindane	Extractable/P		8.25	8.08	HSD
Maleic anhydride	Extractable/BN		8.25	8.06	ECD, FID
Methanol	Volatile		8.24	8.01	FID
Methomyl	Extractable/BN	8.32 (HPLC)			
Methyl ethyl ketone	Volatile		8.25	8.01	FID
				8.02	FID
Methyl isobutyl ketone	Volatile		8.25	8.01	FID
				8.02	FID
Naphthalene	Extractable/BN		8.25	8.10	FID
Napthoquinone	Extractable/BN		8.25	8.06	ECD, FID
				8.09	FID
Nitrobenzene	Extractable/BN		8.25	8.09	ECD, FID
4-Nitrophenol	Extractable/A		8.24	8.04	ECD, FID
Paraldehyde (trimer of acetalde- hyde).	Volatile		8.24	8.01	FID
Pentachlorophenol	Extractable/A		8.25	8.04	ECD
Phenol	Extractable/A		8.25	8.04	ECD, FID
Phorate	Extractable/BN			8.22	FPD
Phosphorodithioic acid esters	Extractable/BN			8.06	ECD, FID
				8.09	ECD, FID
				8.22	FPD
Phthalic anhydride	Extractable/BN		8.25	8.06	ECD, FID
				8.09	ECD, FID
2-Picoline	Extractable/BN		8.25	8.06	ECD, FID
				8.09	ECD, FID
Pyridine	Extractable/BN		8.25	8.06	ECD, FID
				8.09	ECD, FID
Tetrachlorobenzene(s)	Extractable/BN		8.25	8.12	HSD
Tetrachloroethane(s)	Volatile		8.24	8.01	HSD
Tetrachloroethene	Volatile		8.24	8.01	HSD
Tetrachlorophenol	Extractable/A		8.24	8.04	ECD
Toluene	Volatile		8.24	8.02	PID
Toluenediamine	Extractable/BN		8.25		
Toluene diisocyanate(s)	Extractable/nonaqueous		8.25	8.06	FID
Toxaphene	Extractable/P		8.25	8.08	HSD
Trichloroethane	Volatile		8.24	8.01	HSD
Trichloroethene(s)	Volatile		8.24	8.01	HSD
Trichlorofluoromethane	Volatile		8.24	8.01	HSD
Trichlorophenol(s)	Extractable/A		8.25	8.04	HSD
2,4,5-TP (Silvex)	Extractable/A		8.25	8.40	HSD
Trichloropropane	Volatile		8.24	8.01	HSD
Vinyl chloride	Volatile		8.24	8.01	HSD
Vinylidene chloride	Volatile		8.24	8.01	HSD
Xylene	Volatile		8.24	8.02	PID

[1]Analyze for phenanthrene and carbazole; if these are present in a ratio between 1.4:1 and 5:1, creosote should be considered present.
ECD = Electron capture detector; FID = Flame ionization detector; FPD = Flame photometric detector; HSD = Halide specific detector; HPLC = High pressure liquid chromotography; NSD = Nitrogen-specific detector; PID = Photoionization detector.

TABLE 2—ANALYTICAL CHARACTERISTICS OF INORGANIC SPECIES

Species	Sample handling class	Measurement technique	Method number
Antimony	Digestion	Atomic absorbtion–furnace/flame	8.50
Arsenic	Hydride	Atomic absorbtion–flame	8.51
Barium	Digestion	Atomic absorbtion–furnace/flame	8.52
Cadmium	Digestion	Atomic absorbtion–furnace/flame	8.53
Chromium	Digestion	Atomic absorbtion–furnace/flame	8.54
Cyanides	Hydrolysis	Atomic absorbtion-spectroscopy	8.55
Lead	Digestion	Atomic absorbtion–furnace/flame	8.56
Mercury	Cold Vapor	Atomic absorbtion	8.57
Nickel	Digestion	Atomic absorbtion–furnace/flame	8.58
Selenium	Hydride digestion	Atomic absorbtion–furnace/flame	8.59

TABLE 2—ANALYTICAL CHARACTERISTICS OF INORGANIC SPECIES—Continued

Species	Sample handling class	Measurement technique	Method number
Silver	Digestion	Atomic absorbtion–furnace/flame	8.60

TABLE 3.—SAMPLE PREPARTION/SAMPLE INTRODUCTION TECHNIQUES

Sample handling class	Physical characteristics of waste [1]		
	Fluid	Paste	Solid
Volatile	Purge and trap. Direct injection.	Purge and trap. Headspace	Headspace.
Semivolatile and nonvolatile.	Direct injection. Shake out	Shake out	Shake out. Soxhlet. Sonication.
Inorganic	Direct injection. Digestion .... Hydride	Digestion .... Hydride	Digestion. Hydride.

[1] For purposes of this Table, fluid refers to readily pourable liquids, which may or may not contain suspended particles. Paste-like materials, while fluid in the sense of flowability, can be thought of as being thixotropic or plastic in nature, e.g. paints. Solid materials are those wastes which can be handled without a container (i.e., can be piled up without appreciable sagging).

Procedure and Method Number(s)

Digestion—See appropriate procedure for element of interest.
Direct injection—8.80
Headspace—8.82
Hydride—See appropriate procedure for element of interest.
Purge & Trap—8.83
Shake out—8.84
Sonication—8.85
Soxhlet—8.86

[45 FR 33119, May 19, 1980, as amended at 46 FR 35247, July 7, 1981]

APPENDIX IV—[RESERVED FOR RADIOACTIVE WASTE TEST METHODS]

APPENDIX V—[RESERVED FOR INFECTIOUS WASTE TREATMENT SPECIFICATIONS]

APPENDIX VI—[RESERVED FOR ETIOLOGIC AGENTS]

APPENDIX VII—BASIS FOR LISTING HAZARDOUS WASTE

EPA hazardous waste No.	Hazardous constituents for which listed
F001........	Tetrachloroethylene, methylene chloride trichloroethylene, 1,1,1-trichloroethane, carbon tetrachloride, chlorinated fluorocarbons.
F002........	Tetrachloroethylene, methylene chloride, trichloroethylene, 1,1,1-trichloroethane, chlorobenzene, 1,1,2-trichloro-1,2,2-trifluoroethane, ortho-dichlorobenzene, trichlorofluoromethane.
F003........	N.A.
F004........	Cresols and cresylic acid, nitrobenzene.
F005........	Toluene, methyl ethyl ketone, carbon disulfide, isobutanol, pyridine.
F006........	Cadmium, hexavalent chromium, nickel, cyanide (complexed).
F007........	Cyanide (salts).
F008........	Cyanide (salts).
F009........	Cyanide (salts).
F010........	Cyanide (salts).
F011........	Cyanide (salts).
F012........	Cyanide (complexed).
F019........	Hexavalent chromium, cyanide (complexed).
K001........	Pentachlorophenol, phenol, 2-chlorophenol, p-chloro-m-cresol, 2,4-dimethylphenyl, 2,4-dinitrophenol, trichlorophenols, tetrachlorophenols, 2,4-dinitrophenol, cresosote, chrysene, naphthalene, fluoranthene, benzo(b)fluoranthene, benzo(a)pyrene, indeno(1,2,3-cd)pyrene, benz(a)anthracene, dibenz(a)anthracene, acenaphthalene.
K002........	Hexavalent chromium, lead.
K003........	Hexavalent chromium, lead.
K004........	Hexavalent chromium.
K005........	Hexavalent chromium, lead.
K006........	Hexavalent chromium.
K007........	Cyanide (complexed), hexavalent chromium.
K008........	Hexavalent chromium.
K009........	Chloroform, formaldehyde, methylene chloride, methyl chloride, paraldehyde, formic acid.
K010........	Chloroform, formaldehyde, methylene chloride, methyl chloride, paraldehyde, formic acid, chloroacetaldehyde.
K011........	Acrylonitrile, acetonitrile, hydrocyanic acid.
K013........	Hydrocyanic acid, acrylonitrile, acetonitrile.
K014........	Acetonitrile, acrylamide.
K015........	Benzyl chloride, chlorobenzene, toluene, benzotrichloride.
K016........	Hexachlorobenzene, hexachlorobutadiene, carbon tetrachloride, hexachloroethane, perchloroethylene.
K017........	Epichlorohydrin, chloroethers [bis(chloromethyl) ether and bis (2-chloroethyl) ethers], trichloropropane, dichloropropanols.
K018........	1,2-dichloroethane, trichloroethylene, hexachlorobutadiene, hexachlorobenzene.

EPA hazardous waste No.	Hazardous constituents for which listed
K019.......	Ethylene dichloride, 1,1,1-trichloroethane, 1,1,2-trichloroethane, tetrachloroethanes (1,1,2,2-tetrachloroethane and 1,1,1,2-tetrachloroethane), trichloroethylene, tetrachloroethylene, carbon tetrachloride, chloroform, vinyl chloride, vinylidene chloride.
K020.......	Ethylene dichloride, 1,1,1-trichloroethane, 1,1,2-trichloroethane, tetrachloroethanes (1,1,2,2-tetrachloroethane and 1,1,1,2-tetrachloroethane), trichloroethylene, tetrachloroethylene, carbon tetrachloride, chloroform, vinyl chloride, vinylidene chloride.
K021.......	Antimony, carbon tetrachloride, chloroform.
K022.......	Phenol, tars (polycyclic aromatic hydrocarbons).
K023.......	Phthalic anhydride, maleic anhydride.
K024.......	Phthalic anhydride, 1,4-naphthoquinone.
K025.......	Meta-dinitrobenzene, 2,4-dinitrotoluene.
K026.......	Paraldehyde, pyridines, 2-picoline.
K027.......	Toluene diisocyanate, toluene-2, 4-diamine.
K028.......	1,1,1-trichloroethane, vinyl chloride.
K029.......	1,2-dichloroethane, 1,1,1-trichloroethane, vinyl chloride, vinylidene chloride, chloroform.
K030.......	Hexachlorobenzene, hexachlorobutadiene, hexachloroethane, 1,1,1,2-tetrachloroethane, 1,1,2,2-tetrachloroethane, ethylene dichloride.
K031.......	Arsenic.
K032.......	Hexachlorocyclopentadiene.
K033.......	Hexachlorocyclopentadiene.
K034.......	Hexachlorocyclopentadiene.
K035.......	Creosote, chrysene, naphthalene, fluoranthene benzo(b) fluoranthene, benzo(a)pyrene, indeno(1,2,3-cd) pyrene, benzo(a)anthracene, dibenzo(a)anthracene, acenaphthalene.
K036.......	Toluene, phosphorodithioic and phosphorothioic acid esters.
K037.......	Toluene, phosphorodithioic and phosphorothioic acid esters.
K038.......	Phorate, formaldehyde, phosphorodithioic and phosphorothioic acid esters.
K039.......	Phosphorodithioic and phosphorothioic acid esters.
K040.......	Phorate, formaldehyde, phosphorodithioic and phosphorothioic acid esters.
K041.......	Toxaphene.
K042.......	Hexachlorobenzene, ortho-dichlorobenzene.
K043.......	2,4-dichlorophenol, 2,6-dichlorophenol, 2,4,6-trichlorophenol.
K044.......	N.A.
K045.......	N.A.
K046.......	Lead.
K047.......	N.A.
K048.......	Hexavalent chromium, lead.
K049.......	Hexavalent chromium, lead.
K050.......	Hexavalent chromium.
K051.......	Hexavalent chromium, lead.
K052.......	Lead.
K060.......	Cyanide, napthalene, phenolic compounds, arsenic.
K061.......	Hexavalent chromium, lead, cadmium.
K062.......	Hexavalent chromium, lead.
K069.......	Hexavalent chromium, lead, cadmium.
K071.......	Mercury.
K073.......	Chloroform, carbon tetrachloride, hexacholroethane, trichloroethane, tetrachloroethylene, dichloroethylene, 1,1,2,2-tetrachloroethane.
K083.......	Aniline, diphenylamine, nitrobenzene, phenylenediamine.
K084.......	Arsenic.

EPA hazardous waste No.	Hazardous constituents for which listed
K085.......	Benzene, dichlorobenzenes, trichlorobenzenes, tetrachlorobenzenes, pentachlorobenzene, hexachlorobenzene, benzyl chloride.
K086.......	Lead, hexavalent chromium.
K087.......	Phenol, naphthalene.
K093.......	Phthalic anhydride, maleic anhydride.
K094.......	Phthalic anhydride.
K095.......	1,1,2-trichloroethane, 1,1,1,2-tetrachloroethane, 1,1,2,2-tetrachloroethane.
K096.......	1,2-dichloroethane, 1,1,1-trichloroethane, 1,1,2-trichloroethane.
K097.......	Chlordane, heptachlor.
K098.......	Toxaphene.
K099.......	2,4-dichlorophenol, 2,4,6-trichlorophenol.
K100.......	Hexavalent chromium, lead, cadmium.
K101.......	Arsenic.
K102.......	Arsenic.
K103.......	Aniline, nitrobenzene, phenylenediamine.
K104.......	Aniline, benzene, diphenylamine, nitrobenzene, phenylenediamine.
K105.......	Benzene, monochlorobenzene, dichlorobenzenes, 2,4,6-trichlorophenol.
K106.......	Mercury.

N.A.—Waste is hazardous because it fails the test for the characteristic of ignitability, corrosivity, or reactivity.

[46 FR 4619, Jan. 16, 1981, as amended at 46 FR 27477, May 20, 1981]

APPENDIX VIII—HAZARDOUS CONSTITUENTS

Acetonitrile (Ethanenitrile)
Acetophenone (Ethanone, 1-phenyl)
3-(alpha-Acetonylbenzyl)-4-hydroxycoumarin and salts (Warfarin)
2-Acetylaminofluorene (Acetamide, N-(9H-fluoren-2-yl)-)
Acetyl chloride (Ethanoyl chloride)
1-Acetyl-2-thiourea (Acetamide, N-(aminothioxomethyl)-)
Acrolein (2-Propenal)
Acrylamide (2-Propenamide)
Acrylonitrile (2-Propenenitrile)
Aflatoxins
Aldrin   (1,2,3,4,10,10-Hexachloro-1,4,4a,5,8,8a,8b-hexahydro-endo,exo-1,4:5,8-Dimethanonaphthalene)
Allyl alcohol (2-Propen-1-ol)
Aluminum phosphide
4-Aminobiphenyl ([1,1'-Biphenyl]-4-amine)
6-Amino-1,1a,2,8,8a,8b-hexahydro-8-(hydroxymethyl)-8a-methoxy-5-methyl-carbamate azirino[2',3':3,4]pyrrolo[1,2-a]indole-4,7-dione, (ester) (Mitomycin C) (Azirino[2'3':3,4]pyrrolo(1,2-a)indole-4,7-dione,   6-amino-8-[((aminocarbonyl)oxy)methyl]-1,1a,2,8,8a,8b-hexahydro-8amethoxy-5-methy-)
5-(Aminomethyl)-3-isoxazolol   (3(2H)-Isoxazolone, 5-(aminomethyl)-) 4-Aminopyridine (4-Pyridinamine)

Chapter I—Environmental Protection Agency

Amitrole (1H-1,2,4-Triazol-3-amine)
Aniline (Benzenamine)
Antimony and compounds, N.O.S.*
Aramite (Sulfurous acid, 2-chloroethyl-, 2-[4-(1,1-dimethylethyl)phenoxy]-1-methylethyl ester)
Arsenic and compounds, N.O.S.*
Arsenic acid (Orthoarsenic acid)
Arsenic pentoxide (Arsenic (V) oxide)
Arsenic trioxide (Arsenic (III) oxide)
Auramine (Benzenamine, 4,4'-carbonimidoylbis[N,N-Dimethyl-, mono-hydrochloride)
Azaserine (L-Serine, diazoacetate (ester))
Barium and compounds, N.O.S.*
Barium cyanide
Benz[c]acridine (3,4-Benzacridine)
Benz[a]anthracene (1,2-Benzanthracene)
Benzene (Cyclohexatriene)
Benzenearsonic acid (Arsonic acid, phenyl-)
Benzene, dichloromethyl- (Benzal chloride)
Benzenethiol (Thiophenol)
Benzidine ([1,1'-Biphenyl]-4,4'diamine)
Benzo[b]fluoranthene (2,3-Benzofluoranthene)
Benzo[j]fluoranthene (7,8-Benzofluoranthene)
Benzo[a]pyrene (3,4-Benzopyrene)
p-Benzoquinone (1,4-Cyclohexadienedione)
Benzotrichloride (Benzene, trichloromethyl-)
Benzyl chloride (Benzene, (chloromethyl)-)
Beryllium and compounds, N.O.S.*
Bis(2-chloroethoxy)methane (Ethane, 1,1'-[methylenebis(oxy)]bis[2-chloro-])
Bis(2-chloroethyl) ether (Ethane, 1,1'-oxybis[2-chloro-])
N,N-Bis(2-chloroethyl)-2-naphthylamine (Chlornaphazine)
Bis(2-chloroisopropyl) ether (Propane, 2,2'-oxybis[2-chloro-])
Bis(chloromethyl) ether (Methane, oxybis[chloro-])
Bis(2-ethylhexyl) phthalate (1,2-Benzenedicarboxylic acid, bis(2-ethylhexyl) ester)
Bromoacetone (2-Propanone, 1-bromo-)
Bromomethane (Methyl bromide)
4-Bromophenyl phenyl ether (Benzene, 1-bromo-4-phenoxy-)
Brucine (Strychnidin-10-one, 2,3-dimethoxy-)
2-Butanone peroxide (Methyl ethyl ketone, peroxide)
Butyl benzyl phthalate (1,2-Benzenedicarboxylic acid, butyl phenyl-methyl ester)
2-sec-Butyl-4,6-dinitrophenol (DNBP) (Phenol, 2,4-dinitro-6-(1-methylpropyl)-)
Cadmium and compounds, N.O.S.*
Calcium chromate (Chromic acid, calcium salt)

---

* The abbreviation N.O.S. (not otherwise specified) signifies those members of the general class not specifically listed by name in this appendix.

Calcium cyanide
Carbon disulfide (Carbon bisulfide)
Carbon oxyfluoride (Carbonyl fluoride)
Chloral (Acetaldehyde, trichloro-)
Chlorambucil (Butanoic acid, 4-[bis(2-chloroethyl)amino]benzene-)
Chlordane (alpha and gamma isomers) (4,7-Methanoindan, 1,2,4,5,6,7,8,8-octachloro-3,4,7,7a-tetrahydro-) (alpha and gamma isomers)
Chlorinated benzenes, N.O.S.*
Chlorinated ethane, N.O.S.*
Chlorinated fluorocarbons, N.O.S.*
Chlorinated naphthalene, N.O.S.*
Chlorinated phenol, N.O.S.*
Chloroacetaldehyde (Acetaldehyde, chloro-)
Chloroalkyl ethers, N.O.S.*
p-Chloroaniline (Benzenamine, 4-chloro-)
Chlorobenzene (Benzene, chloro-)
Chlorobenzilate (Benzeneacetic acid, 4-chloro-alpha-(4-chlorophenyl)-alpha-hydroxy-, ethyl ester)
p-Chloro-m-cresol (Phenol, 4-chloro-3-methyl)
1-Chloro-2,3-epoxypropane (Oxirane, 2-(chloromethyl)-)
2-Chloroethyl vinyl ether (Ethene, (2-chloroethoxy)-)
Chloroform (Methane, trichloro-)
Chloromethane (Methyl chloride)
Chloromethyl methyl ether (Methane, chloromethoxy-)
2-Chloronaphthalene (Naphthalene, beta-chloro-)
2-Chlorophenol (Phenol, o-chloro-)
1-(o-Chlorophenyl)thiourea (Thiourea, (2-chlorophenyl)-)
3-Chloropropionitrile (Propanenitrile, 3-chloro-)
Chromium and compounds, N.O.S.*
Chrysene (1,2-Benzphenanthrene)
Citrus red No. 2 (2-Naphthol, 1-[(2,5-dimethoxyphenyl)azo]-)
Coal tars
Copper cyanide
Creosote (Creosote, wood)
Cresols (Cresylic acid) (Phenol, methyl-)
Crotonaldehyde (2-Butenal)
Cyanides (soluble salts and complexes), N.O.S.*
Cyanogen (Ethanedinitrile)
Cyanogen bromide (Bromine cyanide)
Cyanogen chloride (Chlorine cyanide)
Cycasin (beta-D-Glucopyranoside, (methyl-ONN-azoxy)methyl-)
2-Cyclohexyl-4,6-dinitrophenol (Phenol, 2-cyclohexyl-4,6-dinitro-)
Cyclophosphamide (2H-1,3,2,-Oxazaphosphorine, [bis(2-chloroethyl)amino]-tetrahydro-, 2-oxide)
Daunomycin (5,12-Naphthacenedione, (8S-cis)-8-acetyl-10-[(3-amino-2,3,6-trideoxy)-alpha-L-lyxo-hexopyranosyl)oxy]-7,8,9,10-tetrahydro-6,8,11-trihydroxy-1-methoxy-)

DDD (Dichlorodiphenyldichloroethane)
(Ethane, 1,1-dichloro-2,2-bis(p-chloro-phenyl)-)
DDE (Ethylene, 1,1-dichloro-2,2-bis(4-chlorophenyl)-)
DDT (Dichlorodiphenyltrichloroethane)
(Ethane, 1,1,1-trichloro-2,2-bis(p-chloro-phenyl)-)
Diallate (S-(2,3-dichloroallyl) diisopropylthiocarbamate)
Dibenz[a,h]acridine (1,2,5,6-Dibenzacridine)
Dibenz[a,j]acridine (1,2,7,8-Dibenzacridine)
Dibenz[a,h]anthracene (1,2,5,6-Dibenzanthracene)
7H-Dibenzo[c,g]carbazole (3,4,5,6-Dibenzcarbazole)
Dibenzo[a,e]pyrene (1,2,4,5-Dibenzpyrene)
Dibenzo[a,h]pyrene (1,2,5,6-Dibenzpyrene)
Dibenzo[a,i]pyrene (1,2,7,8-Dibenzpyrene)
1,2-Dibromo-3-chloropropane (Propane, 1,2-dibromo-3-chloro-)
1,2-Dibromoethane (Ethylene dibromide)
Dibromomethane (Methylene bromide)
Di-n-butyl phthalate (1,2-Benzenedicarboxylic acid, dibutyl ester)
o-Dichlorobenzene (Benzene, 1,2-dichloro-)
m-Dichlorobenzene (Benzene, 1,3-dichloro-)
p-Dichlorobenzene (Benzene, 1,4-dichloro-)
Dichlorobenzene, N.O.S.* (Benzene, dichloro-, N.O.S.*)
3,3'-Dichlorobenzidine ([1,1'-Biphenyl]-4,4'-diamine, 3,3'-dichloro-)
1,4-Dichloro-2-butene (2-Butene, 1,4-dichloro-)
Dichlorodifluoromethane (Methane, dichlorodifluoro-)
1,1-Dichloroethane (Ethylidene dichloride)
1,2-Dichloroethane (Ethylene dichloride)
trans-1,2-Dichloroethene (1,2-Dichloroethylene)
Dichloroethylene, N.O.S.* (Ethene, dichloro-, N.O.S.*)
1,1-Dichloroethylene (Ethene, 1,1-dichloro-)
Dichloromethane (Methylene chloride)
2,4-Dichlorophenol (Phenol, 2,4-dichloro-)
2,6-Dichlorophenol (Phenol, 2,6-dichloro-)
2,4-Dichlorophenoxyacetic acid (2,4-D), salts and esters (Acetic acid, 2,4-dichlorophenoxy-, salts and esters)
Dichlorophenylarsine (Phenyl dichloroarsine)
Dichloropropane, N.O.S.* (Propane, dichloro-, N.O.S.*)
1,2-Dichloropropane (Propylene dichloride)
Dichloropropanol, N.O.S.* (Propanol, dichloro-, N.O.S.*)
Dichloropropene, N.O.S.* (Propene, dichloro-, N.O.S.*)
1,3-Dichloropropene (1-Propene, 1,3-dichloro-)
Dieldrin (1,2,3,4,10.10-hexachloro-6,7-epoxy-1,4,4a,5,6,7,8,8a-octa-hydro-endo,exo-1,4:5,8-Dimethanonaphthalene)
1,2:3,4-Diepoxybutane (2,2'-Bioxirane)
Diethylarsine (Arsine, diethyl-)
N,N-Diethylhydrazine (Hydrazine, 1,2-diethyl)

O,O-Diethyl S-methyl ester of phosphorodithioic acid (Phosphorodithioic acid, O,O-diethyl S-methyl ester)
O,O-Diethylphosphoric acid, O-p-nitrophenyl ester (Phosphoric acid, diethyl p-nitrophenyl ester)
Diethyl phthalate (1,2-Benzenedicarboxylic acid, diethyl ester)
O,O-Diethyl O-2-pyrazinyl phosphorothioate (Phosphorothioic acid, O,O-diethyl O-pyrazinyl ester)
Diethylstilbesterol (4,4'-Stilbenediol, alpha,alpha-diethyl, bis(dihydrogen phosphate, (E)-)
Dihydrosafrole (Benzene, 1,2-methylenedioxy-4-propyl-)
3,4-Dihydroxy-alpha-(methylamino)methyl benzyl alcohol (1,2-Benzenediol, 4-[1-hydroxy-2-(methylamino)ethyl]-)
Diisopropylfluorophosphate (DFP) (Phosphorofluoridic acid, bis(1-methylethyl) ester)
Dimethoate (Phosphorodithioic acid, O,O-dimethyl S-[2-(methylamino)-2-oxoethyl] ester
3,3'-Dimethoxybenzidine ([1,1'-Biphenyl]-4,4'diamine, 3,3'-dimethoxy-)
p-Dimethylaminoazobenzene (Benzenamine, N,N-dimethyl-4-(phenylazo)-)
7,12-Dimethylbenz[a]anthracene (1,2-Benzanthracene, 7,12-dimethyl-)
3,3'-Dimethylbenzidine ([1,1'-Biphenyl]-4,4'-diamine, 3,3'-dimethyl-)
Dimethylcarbamoyl chloride (Carbamoyl chloride, dimethyl-)
1,1-Dimethylhydrazine (Hydrazine, 1,1-dimethyl-)
1,2-Dimethylhydrazine (Hydrazine, 1,2-dimethyl-)
3,3-Dimethyl-1-(methylthio)-2-butanone, O-[(methylamino) carbonyl]oxime (Thiofanox)
alpha,alpha-Dimethylphenethylamine (Ethanamine, 1,1-dimethyl-2-phenyl-)
2,4-Dimethylphenol (Phenol, 2,4-dimethyl-)
Dimethyl phthalate (1,2-Benzenedicarboxylic acid, dimethyl ester)
Dimethyl sulfate (Sulfuric acid, dimethyl ester)
Dinitrobenzene, N.O.S.* (Benzene, dinitro-, N.O.S.*)
4,6-Dinitro-o-cresol and salts (Phenol, 2,4-dinitro-6-methyl-, and salts)
2,4-Dinitrophenol (Phenol, 2,4-dinitro-)
2,4-Dinitrotoluene (Benzene, 1-methyl-2,4-dinitro-)
2,6-Dinitrotoluene (Benzene, 1-methyl-2,6-dinitro-)
Di-n-octyl phthalate (1,2-Benzenedicarboxylic acid, dioctyl ester)
1,4-Dioxane (1,4-Diethylene oxide)
Diphenylamine (Benzenamine, N-phenyl-)
1,2-Diphenylhydrazine (Hydrazine, 1,2-diphenyl-)
Di-n-propylnitrosamine (N-Nitroso-di-n-propylamine)

Disulfoton (O,O-diethyl S-[2-(ethylthio)ethyl] phosphorodithioate)

2,4-Dithiobiuret (Thioimidodicarbonic diamide)

Endosulfan (5-Norbornene, 2,3-dimethanol, 1,4,5,6,7,7-hexachloro-, cyclic sulfite)

Endrin and metabolites (1,2,3,4,10,10-hexachloro-6,7-epoxy-1,4,4a,5,6,7,8,8a-octahydro-endo,endo-1,4:5,8-dimethanonaphthalene, and metabolites)

Ethyl carbamate (Urethan) (Carbamic acid, ethyl ester)

Ethyl cyanide (propanenitrile)

Ethylenebisdithiocarbamic acid, salts and esters (1,2-Ethanediylbiscarbamodithioic acid, salts and esters)

Ethyleneimine (Aziridine)

Ethylene oxide (Oxirane)

Ethylenethiourea (2-Imidazolidinethione)

Ethyl methacrylate (2-Propenoic acid, 2-methyl-, ethyl ester)

Ethyl methanesulfonate (Methanesulfonic acid, ethyl ester)

Fluoranthene (Benzo[j,k]fluorene)

Fluorine

2-Fluoroacetamide (Acetamide, 2-fluoro-)

Fluoroacetic acid, sodium salt (Acetic acid, fluoro-, sodium salt)

Formaldehyde (Methylene oxide)

Formic acid (Methanoic acid)

Glycidylaldehyde (1-Propanol-2,3-epoxy)

Halomethane, N.O.S.*

Heptachlor (4,7-Methano-1H-indene, 1,4,5,6,7,8,8-heptachloro-3a,4,7,7a-tetrahydro-)

Heptachlor epoxide (alpha, beta, and gamma isomers) (4,7-Methano-1H-indene, 1,4,5,6,7,8,8-heptachloro-2,3-epoxy-3a,4,7,7-tetrahydro-, alpha, beta, and gamma isomers)

Hexachlorobenzene (Benzene, hexachloro-)

Hexachlorobutadiene (1,3-Butadiene, 1,1,2,3,4,4-hexachloro-)

Hexachlorocyclohexane (all isomers) (Lindane and isomers)

Hexachlorocyclopentadiene (1,3-Cyclopentadiene, 1,2,3,4,5,5-hexachloro-)

Hexachloroethane (Ethane, 1,1,1,2,2,2-hexachloro-)

1,2,3,4,10,10-Hexachloro-1,4,4a,5,8,8a-hexahydro-1,4:5,8-endo,endo-dimethanonaphthalene (Hexachlorohexahydro-endo,endo-dimethanonaphthalene)

Hexachlorophene (2,2'-Methylenebis(3,4,6-trichlorophenol))

Hexachloropropene (1-Propene, 1,1,2,3,3,3-hexachloro-)

Hexaethyl tetraphosphate (Tetraphosphoric acid, hexaethyl ester)

Hydrazine (Diamine)

Hydrocyanic acid (Hydrogen cyanide)

Hydrofluoric acid (Hydrogen fluoride)

Hydrogen sulfide (Sulfur hydride)

Hydroxydimethylarsine oxide (Cacodylic acid)

Indeno(1,2,3-cd)pyrene (1,10-(1,2-phenylene)pyrene)

Iodomethane (Methyl iodide)

Iron dextran (Ferric dextran)

Isocyanic acid, methyl ester (Methyl isocyanate)

Isobutyl alcohol (1-Propanol, 2-methyl-)

Isosafrole (Benzene, 1,2-methylenedioxy-4-allyl-)

Kepone (Decachlorooctahydro-1,3,4-Methano-2H-cyclobuta[cd]pentalen-2-one)

Lasiocarpine (2-Butenoic acid, 2-methyl-, 7-[(2,3-dihydroxy-2-(1-methoxyethyl)-3-methyl-1-oxobutoxy)methyl]-2,3,5,7a-tetrahydro-1H-pyrrolizin-1-yl ester)

Lead and compounds, N.O.S.*

Lead acetate (Acetic acid, lead salt)

Lead phosphate (Phosphoric acid, lead salt)

Lead subacetate (Lead, bis(acetato-O)tetrahydroxytri-)

Maleic anhydride (2,5-Furandione)

Maleic hydrazide (1,2-Dihydro-3,6-pyridazinedione)

Malononitrile (Propanedinitrile)

Melphalan (Alanine, 3-[p-bis(2-chloroethyl)amino]phenyl-, L-)

Mercury fulminate (Fulminic acid, mercury salt)

Mercury and compounds, N.O.S.*

Methacrylonitrile (2-Propenenitrile, 2-methyl-)

Methanethiol (Thiomethanol)

Methapyrilene (Pyridine, 2-[(2-dimethylamino)ethyl]-2-thenylamino-)

Metholmyl (Acetimidic acid, N-[(methylcarbamoyl)oxy]thio-, methyl ester)

Methoxychlor (Ethane, 1,1,1-trichloro-2,2'-bis(p-methoxyphenyl)-)

2-Methylaziridine (1,2-Propylenimine)

3-Methylcholanthrene (Benz[j]aceanthrylene, 1,2-dihydro-3-methyl-)

Methyl chlorocarbonate (Carbonochloridic acid, methyl ester)

4,4'-Methylenebis(2-chloroaniline) (Benzenamine, 4,4'-methylenebis-(2-chloro-)

Methyl ethyl ketone (MEK) (2-Butanone)

Methyl hydrazine (Hydrazine, methyl-)

2-Methyllactonitrile (Propanenitrile, 2-hydroxy-2-methyl-)

Methyl methacrylate (2-Propenoic acid, 2-methyl-, methyl ester)

Methyl methanesulfonate (Methanesulfonic acid, methyl ester)

2-Methyl-2-(methylthio)propionaldehyde-o-(methylcarbonyl) oxime (Propanal, 2-methyl-2-(methylthio)-, O-[(methylamino)carbonyl]oxime)

N-Methyl-N'-nitro-N-nitrosoguanidine (Guanidine, N-nitroso-N-methyl-N'-nitro-)

Methyl parathion (O,O-dimethyl O-(4-nitrophenyl) phosphorothioate)

Methylthiouracil (4-1H-Pyrimidinone, 2,3-dihydro-6-methyl-2-thioxo-)

Mustard gas (Sulfide, bis(2-chloroethyl)-)

Part 261, App. VIII | Title 40—Protection of Environment

Naphthalene
1,4-Naphthoquinone   (1,4-Naphthalene-
dione)
1-Naphthylamine (alpha-Naphthylamine)
2-Naphthylamine (beta-Naphthylamine)
1-Naphthyl-2-thiourea (Thiourea, 1-naphth-
alenyl-)
Nickel and compounds, N.O.S.*
Nickel carbonyl (Nickel tetracarbonyl)
Nickel cyanide (Nickel (II) cyanide)
Nicotine and salts (Pyridine, (S)-3-(1-
methyl-2-pyrrolidinyl)-, and salts)
Nitric oxide (Nitrogen (II) oxide)
p-Nitroaniline (Benzenamine, 4-nitro-)
Nitrobenzine (Benzene, nitro-)
Nitrogen dioxide (Nitrogen (IV) oxide)
Nitrogen mustard and hydrochloride salt
(Ethanamine, 2-chloro-, N-(2-chloroethyl)-
N-methyl-, and hydrochloride salt)
Nitrogen mustard N-Oxide and hydrochlo-
ride salt (Ethanamine, 2-chloro-, N-(2-
chloroethyl)-N-methyl-, and hydrochlo-
ride salt)
Nitroglycerine   (1,2,3-Propanetriol,   trini-
trate)
4-Nitrophenol (Phenol, 4-nitro-)
4-Nitroquinoline-1-oxide (Quinoline, 4-nitro-
1-oxide-)
Nitrosamine, N.O.S.*
N-Nitrosodi-n-butylamine   (1-Butanamine,
N-butyl-N-nitroso-)
N-Nitrosodiethanolamine (Ethanol,   2,2'-
(nitrosoimino)bis-)
N-Nitrosodiethylamine   (Ethanamine,   N-
ethyl-N-nitroso-)
N-Nitrosodimethylamine (Dimethylnitrosa-
mine)
N-Nitroso-N-ethylurea (Carbamide, N-ethyl-
N-nitroso-)
N-Nitrosomethylethylamine   (Ethanamine,
N-methyl-N-nitroso-)
N-Nitroso-N-methylurea   (Carbamide,   N-
methyl-N-nitroso-)
N-Nitroso-N-methylurethane   (Carbamic
acid, methylnitroso-, ethyl ester)
N-Nitrosomethylvinylamine   (Ethenamine,
N-methyl-N-nitroso-)
N-Nitrosomorpholine (Morpholine, N-ni-
troso-)
N-Nitrosonornicotine (Nornicotine, N-
nitroso-)
N-Nitrosopiperidine (Pyridine, hexahydro-,
N-nitroso-)
Nitrosopyrrolidine (Pyrrole, tetrahydro-, N-
nitroso-)
N-Nitrososarcosine (Sarcosine, N-nitroso-)
5-Nitro-o-toluidine (Benzenamine, 2-methyl-
5-nitro-)
Octamethylpyrophosphoramide   (Diphos-
phoramide, octamethyl-)
Osmium tetroxide (Osmium (VIII) oxide)
7-Oxabicyclo[2.2.1]heptane-2,3-dicarboxylic
acid (Endothal)
Paraldehyde   (1,3,5-Trioxane,   2,4,6-tri-
methyl-)
Parathion (Phosphorothioic acid, O,O-
diethyl O-(p-nitrophenyl) ester

Pentachlorobenzene (Benzene, pentachloro-)
Pentachloroethane (Ethane, pentachloro-)
Pentachloronitrobenzene (PCNB) (Benzene,
pentachloronitro-)
Pentachlorophenol (Phenol, pentachloro-)
Phenacetin   (Acetamide,   N-(4-ethoxy-
phenyl)-)
Phenol (Benzene, hydroxy-)
Phenylenediamine (Benzenediamine)
Phenylmercury acetate (Mercury, acetato-
phenyl-)
N-Phenylthiourea (Thiourea, phenyl-)
Phosgene (Carbonyl chloride)
Phosphine (Hydrogen phosphide)
Phosphorodithioic   acid,   O,O-diethyl   S-
[(ethylthio)methyl] ester (Phorate)
Phosphorothioic acid, O,O-dimethyl O-[p-
((dimethylamino)sulfonyl)phenyl]   ester
(Famphur)
Phthalic acid esters, N.O.S.* (Benzene, 1,2-
dicarboxylic acid, esters, N.O.S.*)
Phthalic   anhydride   (1,2-
Benzenedicarboxylic acid anhydride)
2-Picoline (Pyridine, 2-methyl-)
Polychlorinated biphenyl, N.O.S.*
Potassium cyanide
Potassium silver cyanide (Argentate(1-), di-
cyano-, potassium)
Pronamide (3,5-Dichloro-N-(1,1-dimethyl-2-
propynyl)benzamide)
1,3-Propane sultone (1,2-Oxathiolane, 2,2-
dioxide)
n-Propylamine (1-Propanamine)
Propylthiouracil
(Undecamethylenediamine,   N,N'-bis(2-
chlorobenzyl)-, dihydrochloride)
2-Propyn-1-ol (Propargyl alcohol)
Pyridine
Reserpine   (Yohimban-16-carboxylic   acid,
11,17-dimethoxy-18-[(3,4,5-
trimethoxybenzoyl)oxy]-, methyl ester)
Resorcinol (1,3-Benzenediol)
Saccharin and salts (1,2-Benzoisothiazolin-3-
one, 1,1-dioxide, and salts)
Safrole (Benzene, 1,2-methylenedioxy-4-
allyl-)
Selenious acid (Selenium dioxide)
Selenium and compounds, N.O.S.*
Selenium sulfide (Sulfur selenide)
Selenourea (Carbamimidoselenoic acid)
Silver and compounds, N.O.S.*
Silver cyanide
Sodium cyanide
Streptozotocin (D-Glucopyranose, 2-deoxy-
2-(3-methyl-3-nitrosoureido)-)
Strontium sulfide
Strychnine and salts (Strychnidin-10-one,
and salts)
1,2,4,5-Tetrachlorobenzene   (Benzene,
1,2,4,5-tetrachloro-)
2,3,7,8-Tetrachlorodibenzo-p-dioxin (TCDD)
(Dibenzo-p-dioxin, 2,3,7,8-tetrachloro-)
Tetrachloroethane, N.O.S.* (Ethane, te-
trachloro-, N.O.S.*)

1,1,1,2-Tetrachlorethane (Ethane, 1,1,1,2-tetrachloro-)
1,1,2,2-Tetrachlorethane (Ethane, 1,1,2,2-tetrachloro-)
Tetrachloroethane (Ethene, 1,1,2,2-tetrachloro-)
Tetrachloromethane (Carbon tetrachloride)
2,3,4,6,-Tetrachlorophenol (Phenol, 2,3,4,6-tetrachloro-)
Tetraethyldithiopyrophosphate (Dithiopyrophosphoric acid, tetraethyl-ester)
Tetraethyl lead (Plumbane, tetraethyl-)
Tetraethylpyrophosphate (Pyrophosphoric acide, tetraethyl ester)
Tetranitromethane (Methane, tetranitro-)
Thallium and compounds, N.O.S.*
Thallic oxide (Thallium (III) oxide)
Thallium (I) acetate (Acetic acid, thallium (I) salt)
Thallium (I) carbonate (Carbonic acid, dithallium (I) salt)
Thallium (I) chloride
Thallium (I) nitrate (Nitric acid, thallium (I) salt)
Thallium selenite
Thallium (I) sulfate (Sulfuric acid, thallium (I) salt)
Thioacetamide (Ethanethioamide)
Thiosemicarbazide (Hydrazinecarbothioamide)
Thiourea (Carbamide thio-)
Thiuram (Bis(dimethylthiocarbamoyl) disulfide)
Toluene (Benzene, methyl-)
Toluenediamine (Diaminotoluene)
o-Toluidine hydrochloride (Benzenamine, 2-methyl-, hydrochloride)
Tolylene diisocyanate (Benzene, 1,3-diisocyanatomethyl-)
Toxaphene (Camphene, octachloro-)
Tribromomethane (Bromoform)
1,2,4-Trichlorobenzene (Benzene, 1,2,4-trichloro-)

1,1,1-Trichloroethane (Methyl chloroform)
1,1,2-Trichloroethane (Ethane, 1,1,2-trichloro-)
Trichloroethene (Trichloroethylene)
Trichloromethanethiol (Methanethiol, trichloro-)
Trichloromonofluoromethane (Methane, trichlorofluoro-)
2,4,5-Trichlorophenol (Phenol, 2,4,5-trichloro-)
2,4,6-Trichlorophenol (Phenol, 2,4,6-trichloro-)
2,4,5-Trichlorophenoxyacetic acid (2,4,5-T) (Acetic acid, 2,4,5-trichlorophenoxy-)
2,4,5-Trichlorophenoxypropionic acid (2,4,5-TP) (Silvex) (Propionoic acid, 2-(2,4,5-trichlorophenoxy)-)
Trichloropropane, N.O.S.* (Propane, trichloro-, N.O.S.*)
1,2,3-Trichloropropane (Propane, 1,2,3-trichloro-)
O,O,O-Triethyl phosphorothioate (Phosphorothioic acid, O,O,O-triethyl ester)
sym-Trinitrobenzene (Benzene, 1,3,5-trinitro-)
Tris(1-azridinyl) phosphine sulfide (Phosphine sulfide, tris(1-aziridinyl-)
Tris(2,3-dibromopropyl) phosphate (1-Propanol, 2,3-dibromo-, phosphate)
Trypan blue (2,7-Naphthalenedisulfonic acid, 3,3'-[(3,3'-dimethyl(1,1'-biphenyl)-4,4'-diyl)bis(azo)]bis(5-amino-4-hydroxy-, tetrasodium salt)
Uracil mustard (Uracil 5-[bis(2-chloroethyl)amino]-)
Vanadic acid, ammonium salt (ammonium vanadate)
Vanadium pentoxide (Vanadium (V) oxide)
Vinyl chloride (Ethene, chloro-)
Zinc cyanide
Zinc phosphide

[46 FR 27477, May 20, 1981; 46 FR 29708, June 3, 1981]

AUTHORITY: Secs. 1006, 2002, 3002, 3003, 3004, and 3005, Solid Waste Disposal Act, as amended by the Resource Conservation and Recovery Act of 1976, as amended, (RCRA), (42 U.S.C. 6905, 6912, 6922, 6923, 6924, 6925).

SOURCE: 45 FR 33142, May 19, 1980, unless otherwise noted.

Subpart A—General

§ 262.10 Purpose, scope, and applicability.

(a) These regulations establish standards for generators of hazardous waste.

(b) A generator who treats, stores, or disposes of hazardous waste on-site must only comply with the following sections of this Part with respect to that waste: Section 262.11 for determining whether or not he has a hazardous waste, § 262.12 for obtaining an EPA identification number, § 262.34 for accumulation of hazadous waste, § 262.40 (c) and (d) for recordkeeping, § 262.43 for additional reporting and if applicable, § 262.51 for farmers.

(c) Any person who imports hazardous waste into the United States must comply with the standards applicable to generators established in this Part.

(d) A farmer who generates waste pesticides which are hazardous waste and who complies with all of the requirements of § 262.51 is not required to comply with other standards in this Part or 40 CFR Parts 122, 264, or 265 with respect to such pesticides.

(e) A person who generates a hazardous waste as defined by 40 CFR Part 261 is subject to the compliance requirements and penalties prescribed in Section 3008 of the Act if he does not comply with the requirements of this Part.

(f) An owner or operator who initiates a shipment of hazardous waste from a treatment, storage, or disposal facility must comply with the generator standards established in this Part.

[NOTE: The provisions of § 262.34 are applicable to the on-site accumulation of hazardous waste by generators. Therefore, the provisions of § 262.34 only apply to owners or operators who are shipping hazardous waste which they generated at that facility.]

NOTE: A generator who treats, stores, or disposes of hazardous waste on-site must comply with the applicable standards and permit requirements set forth in 40 CFR Parts 264, 265, and 266 and Part 122.

[45 FR 33142, May 19, 1980, as amended at 45 FR 86970, Dec. 31, 1980; 47 FR 1251, Jan. 11, 1982]

§ 262.11 Hazardous waste determination.

A person who generates a solid waste, as defined in 40 CFR 261.2, must determine if that waste is a hazardous waste using the following method:

(a) He should first determine if the waste is excluded from regulation under 40 CFR 261.4.

(b) He must then determine if the waste is listed as a hazardous waste in Subpart D of 40 CFR Part 261.

NOTE: Even if the waste is listed, the generator still has an opportunity under 40 CFR 260.22 to demonstrate to the Administrator that the waste from his particular facility or operation is not a hazardous waste.

(c) If the waste is not listed as a hazardous waste in Subpart D of 40 CFR Part 261, he must determine whether the waste is identified in Subpart C of 40 CFR Part 261 by either:

(1) Testing the waste according to the methods set forth in Subpart C of 40 CFR Part 261, or according to an equivalent method approved by the Administrator under 40 CFR 260.21; or

(2) Applying knowledge of the hazard characteristic of the waste in light of the materials or the processes used.

[45 FR 33142, May 19, 1980, as amended at 45 FR 76624, Nov. 19, 1980]

§ 262.12 EPA identification numbers.

(a) A generator must not treat, store, dispose of, transport, or offer for transportation, hazardous waste without having received an EPA identification number from the Administrator.

(b) A generator who has not received an EPA identification number may obtain one by applying to the Administrator using EPA form 8700-12. Upon receiving the request the Administrator will assign an EPA identification number to the generator.

(c) A generator must not offer his hazardous waste to transporters or to treatment, storage, or disposal facilities that have not received an EPA identification number.

Subpart B—The Manifest

§ 262.20 General requirements.

(a) A generator who transports, or offers for transportation, hazardous waste for off-site treatment, storage, or disposal must prepare a manifest before transporting the waste off-site.

(b) A generator must designate on the manifest one facility which is permitted to handle the waste described on the manifest.

(c) A generator may also designate on the manifest one alternate facility which is permitted to handle his waste in the event an emergency prevents delivery of the waste to the primary designated facility.

(d) If the transporter is unable to deliver the hazardous waste to the designated facility or the alternate facility, the generator must either designate another facility or instruct the transporter to return the waste.

§ 262.21 Required information.

(a) The manifest must contain all of the following information:

(1) A manifest document number;

(2) The generator's name, mailing address, telephone number, and EPA identification number;

(3) The name and EPA identification number of each transporter;

(4) The name, address and EPA identification number of the designated facility and an alternate facility, if any;

(5) The description of the waste(s) (e.g., proper shipping name, etc.) required by regulations of the U.S. Department of Transportation in 49 CFR 172.101, 172.202, and 172.203;

(6) The total quantity of each hazardous waste by units of weight or volume, and the type and number of containers as loaded into or onto the transport vehicle.

(b) The following certification must appear on the manifest: "This is to certify that the above named materials are properly classified, described, packaged, marked, and labeled and are in proper condition for transportation according to the applicable regulations of the Department of Transportation and the EPA."

§ 262.22 Number of copies.

The manifest consists of at least the number of copies which will provide the generator, each transporter, and the owner or operator of the designated facility with one copy each for their records and another copy to be returned to the generator.

§ 262.23 Use of the manifest.

(a) The generator must:

(1) Sign the manifest certification by hand; and

(2) Obtain the handwritten signature of the initial transporter and date of acceptance on the manifest; and

(3) Retain one copy, in accordance with § 262.40(a).

(b) The generator must give the transporter the remaining copies of the manifest.

(c) For shipments of hazardous waste within the United States solely by water (bulk shipments only), the generator must send three copies of the manifest dated and signed in accordance with this section to the owner or operator of the designated facility or the last water (bulk shipment) transporter to handle the waste in the United States if exported by water. Copies of the manifest are not required for each transporter.

(d) For rail shipments of hazardous waste within the United States which originate at the site of generation, the generator must send at least three copies of the manifest dated and signed in accordance with this section to:

(i) The next non-rail transporter, if any; or

(ii) The designated facility if transported solely by rail; or

(iii) The last rail transporter to handle the waste in the United States if exported by rail.

NOTE: See § 263.20(e) and (f) for special provisions for rail or water (bulk shipment) transporters.

[45 FR 33142, May 19, 1980, as amended at 45 FR 86973, Dec. 31, 1980]

Subpart C—Pre-Transport Requirements

§ 262.30 Packaging.

Before transporting hazardous waste or offering hazardous waste for transportation off-site, a generator must package the waste in accordance with the applicable Department of Transportation regulations on packaging under 49 CFR Parts 173, 178, and 179.

§ 262.31 Labeling.

Before transporting or offering hazardous waste for transportation off-site, a generator must label each package in accordance with the applicable Department of Transportation regulations on hazardous materials under 49 CFR Part 172.

§ 262.32 Marking.

(a) Before transporting or offering hazardous waste for transportation off-site, a generator must mark each package of hazardous waste in accordance with the applicable Department of Transportation regulations on hazardous materials under 49 CFR Part 172;

(b) Before transporting hazardous waste or offering hazardous waste for transportation off-site, a generator must mark each container of 110 gallons or less used in such transportation with the following words and information displayed in accordance with the requirements of 49 CFR 172.304:

HAZARDOUS WASTE—Federal Law Prohibits Improper Disposal. If found, contact the nearest police or public safety authority or the U.S. Environmental Protection Agency.

Generator's Name and Address ———.
Manifest Document Number ———.

§ 262.33 Placarding.

Before transporting hazardous waste or offering hazardous waste for transportation off-site, a generator must placard or offer the initial transporter the appropriate placards according to Department of Transportation regulations for hazardous materials under 49 CFR Part 172, Subpart F.

§ 262.34 Accumulation time.

(a) A generator may accumulate hazardous waste on-site for 90 days or less without a permit or without having interim status provided that:

(1) The waste is placed in containers and the generator complies with Subpart I of 40 CFR Part 265, or the waste is placed in tanks and the generator complies with Subpart J of 40 CFR Part 265 except § 265.193;

(2) The date upon which each period of accumulation begins is clearly marked and visible for inspection on each container;

(3) While being accumulated on-site, each container and tank is labeled or marked clearly with the words, "Hazardous Waste"; and

(4) The generator complies with the requirements for owners or operators in Subparts C and D in 40 CFR Part 265 and with § 265.16.

(b) A generator who accumulates hazardous waste for more than 90 days is an operator of a storage facility and is subject to the requirements of 40 CFR Parts 264 and 265 and the permit requirements of 40 CFR Parts 122 unless he has been granted an extension to the 90-day period. Such extension may be granted by EPA if hazardous wastes must remain on-site for longer than 90 days due to unforeseen, temporary, and uncontrollable circumstances. An extension of up to 30 days may be granted at the discretion of the Regional Administrator on a case-by-case basis.

[47 FR 1251, Jan. 11, 1982]

Subpart D—Recordkeeping and Reporting

§ 262.40 Recordkeeping.

(a) A generator must keep a copy of each manifest signed in accordance with § 262.23(a) for three years or until he receives a signed copy from the designated facility which received the waste. This signed copy must be retained as a record for at least three years from the date the waste was accepted by the initial transporter.

(b) A generator must keep a copy of each Biennial Report and Exception Report for a period of at least three years from the due date of the report.

(c) A generator must keep records of any test results, waste analyses, or other determinations made in accordance with § 262.11 for at least three years from the date that the waste was last sent to on-site or off-site treatment, storage, or disposal.

(d) The periods or retention referred to in this section are extended automatically during the course of any unresolved enforcement action regarding the regulated activity or as requested by the Administrator.

§ 262.41 Biennial report.*

(a) A generator who ships his hazardous waste off-site must prepare and submit a single copy of a biennial report to the Regional Administrator by March 1 of each even numbered year. The biennial report must be submitted on EPA Form 8700–13 A and must cover generator activities during the previous calendar year, and must include the following information:

(1) The EPA identification number, name, and address of the generator;

(2) The calendar year covered by the report;

(3) The EPA identification number, name, and address for each off-site treatment, storage, or disposal facility to which waste was shipped during the year; for exported shipments, the report must give the name and address of the foreign facility.

(4) The name and EPA identification number of each transporter used during the reporting year.

(5) A description, EPA hazardous waste number (from 40 CFR Part 261, Subpart C or D), DOT hazard class, and quantity of each hazardous waste shipped off-site. This information must be listed by EPA identification number of each off-site facility to which waste was shipped.

(6) The certification signed by the generator or his authorized representative.

(b) Any generator who treats, stores, or disposes of hazardous waste on-site must submit a biennial report covering those wastes in accordance with the

*As amended 28 February 1983.

provisions of 40 CFR Parts 122, 264, 265, and 266.

§ 262.42 Exception reporting.

(a) A generator who does not receive a copy of the manifest with the hand-written signature of the owner or operator of the designated facility within 35 days of the date the waste was accepted by the initial transporter must contact the transporter and/or the owner or operator of the designated facility to determine the status of the hazardous waste.

(b) A generator must submit an Exception Report to the EPA Regional Administrator for the Region in which the generator is located if he has not received a copy of the manifest with the handwritten signature of the owner or operator of the designated facility within 45 days of the date the waste was accepted by the initial transporter. The Exception Report must include:

(1) A legible copy of the manifest for which the generator does not have confirmation of delivery;

(2) A cover letter signed by the generator or his authorized representative explaining the efforts taken to locate the hazardous waste and the results of those efforts.

§ 262.43 Additional reporting.

The Administrator, as he deems necessary under section 2002(a) and section 3002(6) of the Act, may require generators to furnish additional reports concerning the quantities and disposition of wastes identified or listed in 40 CFR Part 261.

Subpart E—Special Conditions

§ 262.50 International shipments.

(a) Any person who exports hazardous waste to a foreign country or imports hazardous waste from a foreign country into the United States must comply with the requirements of this Part and with the special requirements of this section.

(b) When shipping hazardous waste outside the United States, the generator must:

(1) Notify the Administrator in writing four weeks before the initial shipment of hazardous waste to each country in each calendar year;

(i) The waste must be identified by its EPA hazardous waste identification number and its DOT shipping description;

(ii) The name and address of the foreign consignee must be included in this notice;

(iii) These notices must be sent to: Hazardous Waste Export, Division for Oceans and Regulatory Affairs (A-107), United States Environmental Protection Agency, Washington, D.C. 20460.

(2) Require that the foreign consignee confirm the delivery of the waste in the foreign country. A copy of the manifest signed by the foreign consignee may be used for this purpose;

(3) Meet the requirements under § 262.21 for the manifest, except that:

(i) In place of the name, address, and EPA identification number of the designated facility, the name and address of the foreign consignee must be used;

(ii) The generator must identify the point of departure from the United States through which the waste must travel before entering a foreign country.

(c) A generator must file an Exception Report, if:

(1) He has not received a copy of the manifest signed by the transporter

NOTE: This requirement to notify will not be delegated to States authorized under 40 CFR Part 123. Therefore, all generators must notify the Administrator as required above.

stating the date and place of departure from the United States within 45 days from the date it was accepted by the initial transporter; or

(2) Within 90 days from the date the waste was accepted by the initial transporter, the generator has not received written confirmation from the foreign consignee that the hazardous waste was received.

(d) When importing hazardous waste, a person must meet all requirements of § 262.21 for the manifest except that:

(1) In place of the generator's name, address and EPA identification number, the name and address of the foreign generator and the importer's name, address and EPA identification number must be used.

(2) In place of the generator's signature on the certification statement, the U.S. importer or his agent must sign and date the certification and obtain the signature of the initial transporter.

§ 262.51 Farmers.

A farmer disposing of waste pesticides from his own use which are hazardous wastes is not required to comply with the standards in this Part or other standards in 40 CFR Parts 122, 264 or 265 for those wastes provided he triple rinses each emptied pesticide container in accordance with § 261.7(b)(3) and disposes of the pesticide residues on his own farm in a manner consistent with the disposal instructions on the pesticide label.

[45 FR 33142, May 19, 1980, as amended at 45 FR 78529, Nov. 25, 1980]

PART 263—STANDARDS APPLICABLE TO TRANSPORTERS OF HAZARDOUS WASTE

Subpart A—General

Sec.
263.10   Scope.
263.11   EPA identification number.
263.12   Transfer facility requirements.

Subpart B—Compliance With the Manifest System and Recordkeeping

263.20   The manifest system.
263.21   Compliance with the manifest.
263.22   Recordkeeping.

Subpart C—Hazardous Waste Discharges

263.30   Immediate action.
263.31   Discharge clean up.

AUTHORITY: Sec. 2002(a), 3002, 3003, 3004 and 3005 of the Solid Waste Disposal Act as amended by the Resource Conservation and Recovery Act of 1976 and as amended by the Quiet Communities Act of 1978, (42 U.S.C. 6912, 6922, 6923, 6924, 6925).

SOURCE: 45 FR 33151, May 19, 1980, unless otherwise noted.

Subpart A—General

§ 263.10   Scope.

(a) These regulations establish standards which apply to persons transporting hazardous waste within the United States if the transportation requires a manifest under 40 CFR Part 262.

NOTE: The regulations set forth in Parts 262 and 263 establish the responsibilities of generators and transporters of hazardous waste in the handling, transportation, and management of that waste. In these regulations, EPA has expressly adopted certain regulations of the Department of Transportation (DOT) governing the transportation of hazardous materials. These regulations concern, among other things, labeling, marking, placarding, using proper containers, and reporting discharges. EPA has expressly adopted these regulations in order to satisfy its statutory obligation to promulgate regulations which are necessary to protect human health and the environment in the transportation of hazardous waste. EPA's adoption of these DOT regulations ensures consistency with the requirements of DOT and thus avoids the establishment of duplicative or conflicting requirements with respect to these matters. These EPA regulations which apply to both interstate and intrastate transportation of hazardous waste are enforceable by EPA.

DOT has revised its hazardous materials transportation regulations in order to encompass the transportation of hazardous waste and to regulate intrastate, as well as interstate, transportation of hazardous waste. Transporters of hazardous waste are cautioned that DOT's regulations are fully applicable to their activities and enforceable by DOT. These DOT regulations are codified in Title 49, Code of Federal Regulations, Subchapter C.

EPA and DOT worked together to develop standards for transporters of hazardous waste in order to avoid conflicting requirements. Except for transporters of bulk shipments of hazardous waste by water, a transporter who meets all applicable requirements of 49 CFR Parts 171 through 179 and the requirements of 40 CFR 263.11 and 263.31 will be deemed in compliance with this part. Regardless of DOT's action, EPA retains its authority to enforce these regulations.

(b) These regulations do not apply to on-site transportation of hazardous waste by generators or by owners or operators of permitted hazardous waste management facilities.

(c) A transporter of hazardous waste must also comply with 40 CFR Part 262, Standards Applicable to Generators of Hazardous Waste, if he:

(1) Transports hazardous waste into the United States from abroad; or

(2) Mixes hazardous wastes of different DOT shipping descriptions by placing them into a single container.

[45 FR 33151, May 19, 1980, as amended at 45 FR 86968, Dec. 31, 1980]

§ 263.11   EPA identification number.

(a) A transporter must not transport hazardous wastes without having received an EPA identification number from the Administrator.

(b) A transporter who has not received an EPA identification number may obtain one by applying to the Administrator using EPA Form 8700-12. Upon receiving the request, the Administrator will assign an EPA identification number to the transporter.

§ 263.12   Transfer facility requirements.

A transporter who stores manifested shipments of hazardous waste in containers meeting the requirements of § 262.30 at a transfer facility for a period of ten days or less is not subject to regulation under Parts 122, 264, and 265 of this chapter with respect to the storage of those wastes.

[45 FR 86968, Dec. 31, 1980]

Subpart B—Compliance With the Manifest System and Recordkeeping

§ 263.20 The manifest system.

(a) A transporter may not accept hazardous waste from a generator unless it is accompanied by a manifest, signed by the generator in accordance with the provisions of 40 CFR Part 262.

(b) Before transporting the hazardous waste, the transporter must sign and date the manifest acknowledging acceptance of the hazardous waste from the generator. The transporter must return a signed copy to the generator before leaving the generator's property.

(c) The transporter must ensure that the manifest accompanies the hazardous waste.

(d) A transporter who delivers a hazardous waste to another transporter or to the designated facility must:

(1) Obtain the date of delivery and the handwritten signature of that transporter or of the owner or operator of the designated facility on the manifest; and

(2) Retain one copy of the manifest in accordance with § 263.22; and

(3) give the remaining copies of the manifest to the accepting transporter or designated facility.

(e) The requirements of paragraphs (c), (d) and (f) of this section do not apply to water (bulk shipment) transporters if:

(1) The hazardous waste is delivered by water (bulk shipment) to the designated facility; and

(2) A shipping paper containing all the information required on the manifest (excluding the EPA identification numbers, generator certification, and signatures) accompanies the hazardous waste; and

(3) The delivering transporter obtains the date of delivery and handwritten signature of the owner or op-

erator of the designated facility on either the manifest or the shipping paper; and

(4) The person delivering the hazardous waste to the initial water (bulk shipment) transporter obtains the date of delivery and signature of the water (bulk shipment) transporter on the manifest and forwards it to the designated facility; and

(5) A copy of the shipping paper or manifest is retained by each water (bulk shipment) transporter in accordance with § 263.22.

(f) For shipments involving rail transportation, the requirements of paragraphs (c), (d) and (e) do not apply and the following requirements do apply:

(1) When accepting hazardous waste from a non-rail transporter, the initial rail transporter must:

(i) Sign and date the manifest acknowledging acceptance of the hazardous waste;

(ii) Return a signed copy of the manifest to the non-rail transporter;

(iii) Forward at least three copies of the manifest to:

(A) The next non-rail transporter, if any; or,

(B) The designated facility, if the shipment is delivered to that facility by rail; or

(C) The last rail transporter designated to handle the waste in the United States;

(iv) Retain one copy of the manifest and rail shipping paper in accordance with § 263.22.

(2) Rail transporters must ensure that a shipping paper containing all the information required on the manifest (excluding the EPA identification numbers, generator certification, and signatures) accompanies the hazardous waste at all times.

NOTE: Intermediate rail transporters are not required to sign either the manifest or shipping paper.

(3) When delivering hazardous waste to the designated facility, a rail transporter must:

(i) Obtain the date of delivery and handwritten signature of the owner or operator of the designated facility on the manifest or the shipping paper (if the manifest has not been received by the facility); and

(ii) Retain a copy of the manifest or signed shipping paper in accordance with § 263.22.

(4) When delivering hazardous waste to a non-rail transporter a rail transporter must:

(i) Obtain the date of delivery and the handwritten signature of the next non-rail transporter on the manifest; and

(ii) Retain a copy of the manifest in accordance with § 263.22.

(5) Before accepting hazardous waste from a rail transporter, a non-rail transporter must sign and date the manifest and provide a copy to the rail transporter.

(g) Transporters who transport hazardous waste out of the United States must:

(1) Indicate on the manifest the date the hazardous waste left the United States; and

(2) Sign the manifest and retain one copy in accordance with § 263.22(c); and

(3) Return a signed copy of the manifest to the generator.

[45 FR 33151, May 19, 1980, as amended at 45 FR 86973, Dec. 31, 1980]

§ 263.21 Compliance with the manifest.

(a) The transporter must deliver the entire quantity of hazardous waste which he has accepted from a generator or a transporter to:

(1) The designated facility listed on the manifest; or

(2) The alternate designated facility, if the hazardous waste cannot be delivered to the designated facility because an emergency prevents delivery; or

(3) The next designated transporter; or

(4) The place outside the United States designated by the generator.

(b) If the hazardous waste cannot be delivered in accordance with paragraph (a) of this section, the transporter must contact the generator for further directions and must revise the manifest according to the generator's instructions.

§ 263.22 Recordkeeping.

(a) A transporter of hazardous waste must keep a copy of the manifest signed by the generator, himself, and the next designated transporter or the owner or operator of the designated facility for a period of three years from the date the hazardous waste was accepted by the initial transporter.

(b) For shipments delivered to the designated facility by water (bulk shipment), each water (bulk shipment) transporter must retain a copy of the shipping paper containing all the information required in § 263.20(e)(2) for a period of three years from the date the hazardous waste was accepted by the initial transporter.

(c) For shipments of hazardous waste by rail within the United States:

(i) The initial rail transporter must keep a copy of the manifest and shipping paper with all the information required in § 263.20(f)(2) for a period of three years from the date the hazardous waste was accepted by the initial transporter; and

(ii) The final rail transporter must keep a copy of the signed manifest (or the shipping paper if signed by the designated facility in lieu of the manifest) for a period of three years from the date the hazardous waste was accepted by the initial transporter.

NOTE: Intermediate rail transporters are not required to keep records pursuant to these regulations.

(d) A transporter who transports hazardous waste out of the United States must keep a copy of the manifest indicating that the hazardous waste left the United States for a period of three years from the date the hazardous waste was accepted by the initial transporter.

(e) The periods of retention referred to in this Section are extended automatically during the course of any unresolved enforcement action regarding the regulated activity or as requested by the Administrator.

[45 FR 33151, May 19, 1980, as amended at 45 FR 86973, Dec. 31, 1980]

Subpart C—Hazardous Waste Discharges

§ 263.30 **Immediate action.**

(a) In the event of a discharge of hazardous waste during transportation, the transporter must take appropriate immediate action to protect human health and the environment (e.g., notify local authorities, dike the discharge area).

(b) If a discharge of hazardous waste occurs during transportation and an official (State or local government or a Federal Agency) acting within the scope of his official responsibilities determines that immediate removal of the waste is necessary to protect human health or the environment, that official may authorize the removal of the waste by transporters who do not have EPA identification numbers and without the preparation of a manifest.

(c) An air, rail, highway, or water transporter who has discharged hazardous waste must:

(1) Give notice, if required by 49 CFR 171.15, to the National Response Center (800-424-8802 or 202-426-2675); and

(2) Report in writing as required by 49 CFR 171.16 to the Director, Office of Hazardous Materials Regulations, Materials Transportation Bureau, Department of Transportation, Washington, D.C. 20590.

(d) A water (bulk shipment) transporter who has discharged hazardous waste must give the same notice as required by 33 CFR 153.203 for oil and hazardous substances.

§ 263.31 **Discharge clean up.**

A transporter must clean up any hazardous waste discharge that occurs during transportation or take such action as may be required or approved by Federal, State, or local officials so that the hazardous waste discharge no longer presents a hazard to human health or the environment.

§ 265.316 Disposal of small containers of hazardous waste in overpacked drums (lab packs).

Small containers of hazardous waste in overpacked drums (lab packs) may be placed in a landfull if the following requirements are met:

(a) Hazardous waste must be packaged in non-leaking inside containers. The inside containers must be of a design and constructed of a material that will not react dangerously with, be decomposed by, or be ignited by the waste held therein. Inside containers must be tightly and securely sealed. The inside containers must be of the size and type specified in the Department of Transportation (DOT) hazardous materials regulations (49 CFR Parts 173, 178 and 179), if those regulations specify a particular inside container for the waste.

(b) The inside containers must be overpacked in an open head DOT-specification metal shipping container (49 CFR Parts 178 and 179) of no more than 416-liter (110 gallon) capacity and surrounded by, at a minimum, a sufficient quantity of absorbent material to completely absorb all of the liquid contents of the inside containers. The metal outer container must be full after packing with inside containers and absorbent material.

(c) The absorbent material used must not be capable of reacting dangerously with, being decomposed by, or being ignited by the contents of the inside containers, in accordance with § 265.17(b).

(d) Incompatible wastes, as defined in § 260.10(a) of this chapter, must not be placed in the same outside container.

(e) Reactive waste, other than cyanide- or sulfide-bearing waste as defined in § 261.23(a)(5) of this chapter, must be treated or rendered non-reactive prior to packaging in accordance with paragraphs (a) through (d) of this section. Cyanide- and sulfide-bearing reactive waste may be packaged in accordance with paragraphs (a) through (d) of this section without first being treated or rendered non-reactive.

[46 FR 56596, Nov. 17, 1981]

State Hazardous-Waste Agencies and Regulations on Small Quantity Generator Exemption

State	Agency Telephone	Comparison with Federal SQGE[a,b]		
		Similar	More Stringent	No SQGE[a]
Alabama	(205) 277-3630	X		
Alaska	(907) 465-2666	X		
Arizona	(602) 225-1162	X		
Arkansas	(501) 562-7444	X		
California	(916) 322-2337			X
Colorado	(303) 320-8333	X		
Connecticut	(203) 566-5712		X	
Delaware	(302) 736-4781	X		
Florida	(904) 488-0300	X		
Georgia	(404) 656-2833	X		
Hawaii	(808) 548-6767	X		
Idaho	(208) 334-4064	X		
Illinois	(217) 782-6760		X	
Indiana	(317) 633-0176		X	
Iowa	(515) 281-8853		X	
Kansas	(913) 862-9360		X	
Kentucky	(502) 564-6716	X		
Louisiana	(504) 342-1227		X	
Maine	(800) 452-1942	X		
Maryland	(301) 383-5734	X		

State	Agency Telephone	Comparison with Federal SQGE[a,b]		
		Similar	More Stringent	No SQGE[a]
Massachusetts	(617) 292-5500		X	
Michigan	(517) 373-2730		X	
Minnesota	(612) 297-2735			X
Mississippi	(601) 961-5171	X		
Missouri	(314) 751-3241		X	
Montana	(406) 449-2821	X		
Nebraska	(402) 471-2186	X		
Nevada	(702) 885-4670	X		
New Hampshire	(603) 271-4608		X	
New Jersey	(609) 292-8341		X	
New Mexico	(505) 984-0020	X		
New York	(518) 457-3274		X	
North Carolina	(919) 733-2178	X		
North Dakota	(701) 224-2366	X		
Ohio	(614) 466-7220	X		
Oklahoma	(405) 271-5338	X		
Oregon	(503) 229-5913		X	
Pennsylvania	(717) 787-7381	X		
Rhode Island	(401) 277-2797			X
South Carolina	(803) 758-5681	X		
South Dakota	(605) 773-3329	X		
Tennessee	(615) 741-3424		X	
Texas	(512) 458-7271	X		
Utah	(801) 533-4145	X		
Vermont	(802) 828-3395		X	
Virginia	(804) 786-5271	X		
Washington	(206) 459-6301		X	
West Virginia	(304) 348-5935		X	
Wisconsin	(608) 266-1327		X	
Wyoming	(307) 777-7752	X		

[a]Small Quantity Generator Exemption.
[b]The comparison is with state regulations in effect January 1983.

Regulations on Storage of Laboratory Waste

This appendix is intended to provide a guide to the hazardous-waste storage regulations that are important to laboratories that handle modest quantities of waste from a relatively few points of generation. Large multisite organizations, particularly those that handle large quantities of waste and that store it for periods greater than 90 days, will find it necessary to refer to all of 40 CFR Parts 262, 264, and 265.

A summary and copy of 40 CFR 262 appear in Appendix A. Copies of those parts of 40 CFR 265, Interim Status Standards for Owners and Operators of Hazardous Waste Treatment, Storage, and Disposal Facilities, that are pertinent to smaller laboratories are included in this appendix. These regulations are those in effect 1 April 1983; readers must keep abreast of future changes in them (see Appendix M).

SMALL QUANTITY GENERATOR

If an organization qualifies as a Small Quantity Generator (Appendix A, 40 CFR 261.5), its wastes are not subject to EPA storage regulations and no storage permit is required. There are no time limits on waste storage, but the quantity of waste stored must not exceed the limit for the Small Quantity Exemption. These exemptions may not apply in states that do not recognize the EPA Small Quantity Generator Exemption (see Appendix B). However, an exempt generator must treat or dispose of wastes at a facility that has an EPA or state permit.

STORAGE WITHOUT AN EPA PERMIT (Appendix A, 40 CFR 262.34)*

If a generator of waste does not qualify for the Small Quantity Exemption, hazardous wastes may be accumulated on site without a permit for 90 days or less, provided that

1. The waste is shipped off site in 90 days or less. EPA has discretion to grant a 30-day extension for storage. Beyond this period the laboratory site must apply for a permit.
2. The waste is placed in containers of good integrity in an area well maintained, dated when each period of accumulation begins, and labeled "Hazardous Waste."
3. The generator complies with the requirements for an owner and operator of waste-handling facilities, which cover

- Preparedness and Preventive Measures (40 CFR 265 Subpart C).
- Contingency Plan and Emergency Procedures (40 CFR 265 Subpart D).
- Personal Training (40 CFR 265.16).

STORAGE WITH AN EPA PERMIT

A permit must be obtained for storage in excess of 90 days (see Permit Requirements 40 CFR Part 122). The generator is subject to the full requirements of the Standards in 40 CFR Part 265 (interim status) or Part 264. In general, this involves writing and following a series of plans, which include

1. Waste Analysis (40 CFR 265.13).
2. Security (40 CFR 265.14).
3. Inspections (40 CFR 265.15 and Subpart I or J).
4. Training (40 CFR 265.16).
5. Preparedness and Prevention (40 CFR 265 Subpart C).
6. Contingency and Emergency Procedures (40 CFR 265 Subpart D).
7. Manifest system, Recordkeeping, and Reporting (40 CFR 265 Subpart E).

---

*See Appendix A for a proposed amendment to this section.

In addition, the storage facility must meet some minimum requirements.

1. Containers (40 CFR 265 Subpart I). Containers must be made of or lined with materials compatible with the waste, must be kept closed, and must be within a containment area that is capable of holding spills, leaks, and precipitation (40 CFR 265.17).

2. Tanks (40 CFR 265 Subpart J) must be of sufficient strength and made of materials compatible with the waste. They must have controls to prevent overfilling and, in the case of open tanks, must have sufficient freeboard to prevent overtopping.

AUTHORITY: Secs. 1006, 2002(a), and 3004, Solid Waste Disposal Act, as amended by the Resource Conservation and Recovery Act of 1976, as amended (42 U.S.C. 6905, 6912(a), and 6924).

SOURCE: 45 FR 33232, May 19, 1980, unless otherwise noted.

Subpart A—General

§ 265.1 Purpose, scope, and applicability.

(a) The purpose of this part is to establish minimum national standards which define the acceptable management of hazardous waste during the period of interim status.

(b) The standards in this part apply to owners and operators of facilities which treat, store, or dispose of hazardous waste who have fully complied with the requirements for interim status under section 3005(e) of RCRA and § 122.22 of this Chapter, until final administrative disposition of their permit application is made. These standards apply to all treatment, storage, or disposal of hazardous waste at these facilities after the effective date of these regulations, except as specifically provided otherwise in this part or Part 261 of this Chapter.

[*Comment:* As stated in section 3005(a) of RCRA, after the effective date of regulations under that section, i.e., Parts 122 and 124 of this Chapter, the treatment, storage, or disposal of hazardous waste is prohibited except in accordance with a permit. Section 3005(e) of RCRA provides for the continued operation of an existing facility which meets certain conditions until final administrative disposition of the owner's and operator's permit application is made.]

(c) The requirements of this part do not apply to:

(1) A person disposing of hazardous waste by means of ocean disposal subject to a permit issued under the Marine Protection, Research, and Sanctuaries Act;

[*Comment:* These Part 265 regulations do apply to the treatment or storage of hazardous waste before it is loaded onto an ocean vessel for incineration or disposal at sea, as provided in paragraph (b) of this section.]

(2) A person disposing of hazardous waste by means of underground injection subject to a permit issued under an Underground Injection Control (UIC) program approved or promulgated under the Safe Drinking Water Act;

[*Comment:* These Part 265 regulations do apply to the aboveground treatment or storage of hazardous waste before it is injected underground. These Part 265 regulations also apply to the disposal of hazardous waste by means of underground injection, as provided in paragraph (b) of this Section, until final administrative disposition of a person's permit application is made under RCRA or under an approved or promulgated UIC program.]

(3) The owner or operator of a POTW which treats, stores, or disposes of hazardous waste;

[*Comment:* The owner or operator of a facility under paragraphs (c)(1) through (c)(3) of this section is subject to the requirements of Part 264 of this Chapter to the extent they are included in a permit by rule granted to such a person under Part 122 of this Chapter, or are required by § 122.45 of this Chapter.]

(4) A person who treats, stores, or disposes of hazardous waste in a State with a RCRA hazardous waste program authorized under Subparts A and B, or Subpart F, of Part 123 of this Chapter, except that the requirements of this Part will continue to apply as stated in paragraph (c)(2) of this Section, if the authorized State RCRA program does not cover disposal of hazardous waste by means of underground injection;

(5) The owner or operator of a facility permitted, licensed, or registered by a State to manage municipal or industrial solid waste, if the only hazardous waste the facility treats, stores, or disposes of is excluded from regulation under this Part by § 261.5 of this Chapter;

(6) The owner or operator of a facility which treats or stores hazardous waste, which treatment or storage meets the criteria in § 261.6(a) of this Chapter, except to the extent that § 261.6(b) of this Chapter provides otherwise;

(7) A generator accumulating waste on-site in compliance with § 262.34 of this Chapter, except to the extent the requirements are included in § 262.34 of this Chapter;

(8) A farmer disposing of waste pesticides from his own use in compliance with § 262.51 of this Chapter; or

(9) The owner or operator of a totally enclosed treatment facility, as defined in § 260.10.

(10) The owner or operator of an elementary neutralization unit or a wastewater treatment unit as defined in § 260.10 of this chapter.

(11) Persons with respect to those activities which are carried out to immediateely contain or treat a spill of hazardous waste or material which, when spilled, becomes a hazardous waste, except that, with respect to such activities, the appropriate requirements of Subparts C and D of this part are applicable to owners and operators of treatment, storage and disposal facilities otherwise subject ot this part.

[*Comments:* This paragraph only applies to activities taken in immediate response activities are completed, the regulations of this Chapter apply fully to the management of any spill residue or debris which is a hazardous waste under Part 261.]

(12) A transporter storing manifested shipments of hazardous waste in containers meeting the requirements of 40 CFR 262.30 at a transfer facility for a period of ten days or less.

(13) The addition of absorbent material to waste in a container (as defined in § 260.10 of this chapter) or the addition of waste to the absorbent material in a container provided that these actions occur at the time waste is first placed in the containers; and §§ 265.17(b), 265.171, and 265.172 are complied with.

[45 FR 33232, May 19, 1980, as amended at 45 FR 76075, Nov. 17, 1980; 45 FR 76630, Nov. 19, 1980; 45 FR 86968, Dec. 31, 1980; 46 FR 27480, May 20, 1981; 47 FR 8306, Feb. 25, 1982]

§§ 265.2—265.3 [Reserved]

§ 265.4 Imminent hazard action.

Notwithstanding any other provisions of these regulations, enforcement actions may be brought pursuant to section 7003 of RCRA.

Subpart B—General Facility Standards

§ 265.10 Applicability

The regulations in this Subpart apply to owners and operators of all hazardous waste facilities, except as § 265.1 provides otherwise.

§ 265.11 Identification number.

Every facility owner or operator must apply to EPA for an EPA identification number in accordance with the EPA notification procedures (45 FR 12746).

§ 265.12 **Required notices.**

(a) The owner or operator of a facility that has arranged to receive hazardous waste from a foreign source must notify the Regional Administrator in writing at least four weeks in advance of the date of the waste is expected to arrive at the facility. Notice of subsequent shipments of the same waste from the same foreign source is not required.

(b) Before transferring ownership or operation of a facility during its operating life, or of a disposal facility during the post-closure care period, the owner or operator must notify the new owner or operator in writing of the requirements of this part and Part 122 of this Chapter. (Also see § 122.23(c) of this Chapter.)

[*Comment:* An owner's or operator's failure to notify the new owner or operator of the requirements of this part in no way relieves the new owner or operator of his obligation to comply with all applicable requirements.]

§ 265.13 **General waste analysis.**

(a)(1) Before an owner or operator treats, stores, or disposes of any hazardous waste, he must obtain a detailed chemical and physical analysis of a representative sample of the waste. At a minimum, this analysis must contain all the information which must be known to treat, store, or dispose of the waste in accordance with the requirements of this part.

(2) The analysis may include data developed under Part 261 of this Chapter, and existing published or documented data on the hazardous waste or on waste generated from similar processes.

[*Comment:* For example, the facility's record of analyses performed on the waste before the effective date of these regulations, or studies conducted on hazardous waste generated from processes similar to that which generated the waste to be managed at the facility, may be included in the data base required to comply with paragraph (a)(1) of this section. The owner or operator of an off-site facility may arrange for the generator of the hazardous waste to supply part or all of the information required by paragraph (a)(1) of this section. If the generator does not supply the information, and the owner or operator chooses to accept a hazardous waste, the owner or operator is responsible for obtaining the information required to comply with this section.]

(3) The analysis must be repeated as necessary to ensure that it is accurate and up to date. At a minimum, the analysis must be repeated:

(i) When the owner or operator is notified, or has reason to believe, that the process or operation generating the hazardous waste has changed; and

(ii) For off-site facilities, when the results of the inspection required in paragraph (a)(4) of this section indicate that the hazardous waste received at the facility does not match the waste designated on the accompanying manifest or shipping paper.

(4) The owner or operator of an off-site facility must inspect and, if necessary, analyze each hazardous waste movement received at the facility to determine whether it matches the identity of the waste specified on the accompanying manifest or shipping paper.

(b) The owner or operator must develop and follow a written waste analysis plan which describes the procedures which he will carry out to comply with paragraph (a) of this Section. He must keep this plan at the facility. At a minimum, the plan must specify:

(1) The parameters for which each hazardous waste will be analyzed and the rationale for the selection of these parameters (i.e., how analysis for these parameters will provide sufficient information on the waste's properties to comply with paragraph (a) of this Section);

(2) The test methods which will be used to test for these parameters;

(3) The sampling method which will be used to obtain a representative sample of the waste to be analyzed. A representative sample may be obtained using either:

(i) One of the sampling methods described in Appendix I of Part 261 of this Chapter; or

(ii) An equivalent sampling method.

[*Comment:* See § 260.20(c) of this Chapter for related discussion.]

(4) The frequency with which the initial analysis of the waste will be re-

viewed or repeated to ensure that the analysis is accurate and up to date;

(5) For off-site facilities, the waste analyses that hazardous waste generators have agreed to supply; and

(6) Where applicable, the methods which will be used to meet the additional waste analysis requirements for specific waste management methods as specified in §§ 265.193, 265.225, 265.252, 265.273, 265.345, 265.375, and 265.402.

(c) For off-site facilities, the waste analysis plan required in paragraph (b) of this Section must also specify the procedures which will be used to inspect and, if necessary, analyze each movement of hazardous waste received at the facility to ensure that it matches the identity of the waste designated on the accompanying manifest or shipping paper. At a minimum, the plan must describe:

(1) The procedures which will be used to determine the identity of each movement of waste managed at the facility; and

(2) The sampling method which will be used to obtain a representative sample of the waste to be identified, if the identification method includes sampling.

§ 265.14 Security.

(a) The owner or operator must prevent the unknowing entry, and minimize the possibility for the unauthorized entry, of persons or livestock onto the active portion of his facility, *unless:*

(1) Physical contact with the waste, structures, or equipment with the active portion of the facility will not injure unknowing or unauthorized persons or livestock which may enter the active portion of a facility, and

(2) Disturbance of the waste or equipment, by the unknowing or unauthorized entry of persons or livestock onto the active portion of a facility, will not cause a violation of the requirements of this part.

(b) Unless exempt under paragraphs (a)(1) and (a)(2) of this section, a facility must have:

(1) A 24-hour surveillance system (e.g., television monitoring or surveillance by guards of facility personnel) which continuously monitors and con-

trols entry onto the active portion of the facility; or

(2)(i) An artificial or natural barrier (e.g., a fence in good repair or a fence combined with a cliff), which completely surrounds the active portion of the facility; and

(ii) A means to control entry, at all times, through the gates or other entrances to the active portion of the facility (e.g., an attendant, television monitors, locked entrance, or controlled roadway access to the facility).

[*Comment:* The requirements of paragraph (b) of this section are satisfied if the facility or plant within which the active portion is located itself has a surveillance system, or a barrier and a means to control entry, which complies with the requirements of paragraph (b)(1) or (b)(2) of this section.]

(c) Unless exempt under paragraphs (a)(1) and (a)(2) of this section, a sign with the legend, "Danger—Unauthorized Personnel Keep Out," must be posted at each entrance to the active portion of a facility, and at other locations, in sufficient numbers to be seen from any approach to this active portion. The legend must be written in English and in any other language predominant in the area surrounding the facility (e.g., facilities in counties bordering the Canadian province of Quebec must post signs in French; facilities in counties bordering Mexico must post signs in Spanish), and must be legible from a distance of at least 25 feet. Existing signs with a legend other than "Danger—Unauthorized Personnel Keep Out" may be used if the legend on the sign indicates that only authorized personnel are allowed to enter the active portion, and that entry onto the active portion can be dangerous.

[*Comment:* See § 265.117(b) for discussion of security requirements at disposal facilities during the post-closure care period.]

§ 265.15 General inspection requirements.

(a) The owner or operator must inspect his facility for malfunctions and deterioration, operator errors, and discharges which may be causing—or may lead to: (1) Release of hazardous waste constituents to the environment or (2) a threat to human health. The owner or operator must conduct these inspections often enough to identify problems in time to correct them before they harm human health or the environment.

(b)(1) The owner or operator must develop and follow a written schedule for inspecting all monitoring equipment, safety and emergency equipment, security devices, and operating and structural equipment (such as dikes and sump pumps) that are important to preventing, detecting, or responding to environmental or human health hazards.

(2) He must keep this schedule at the facility.

(3) The schedule must identify the types of problems (e.g., malfunctions or deterioration) which are to be looked for during the inspection (e.g., inoperative sump pump, leaking fitting, eroding dike, etc.).

(4) The frequency of inspection may vary for the items on the schedule. However, it should be based on the rate of possible deterioration of the equipment and the probability of an environmental or human health incident if the deterioration or malfunction or any operator error goes undetected between inspections. Areas subject to spills, such as loading and unloading areas, must be inspected daily when in use. At a minimum, the inspection schedule must include the items and frequencies called for in §§ 265.174, 265.194, 265.226, 265.347, 265.377, and 265.403.

(c) The owner or operator must remedy any deterioration or malfunction of equipment or structures which the inspection reveals on a schedule which ensures that the problem does not lead to an environmental or human health hazard. Where a hazard is imminent or has already occurred, remedial action must be taken immediately.

(d) The owner or operator must record inspections in an inspection log or summary. He must keep these records for at least three years from the date of inspection. At a minimum, these records must include the date and time of the inspection, the name of the inspector, a notation of the observations made, and the date and nature of any repairs or other remedial actions.

§ 265.16 **Personnel training.**

(a)(1) Facility personnel must successfully complete a program of classroom instruction or on-the-job training that teaches them to perform their duties in a way that ensures the facility's compliance with the requirements of this part. The owner or operator must ensure that this program includes all the elements described in the document required under paragraph (d)(3) of this section.

(2) This program must be directed by a person trained in hazardous waste management procedures, and must include instruction which teaches facility personnel hazardous waste management procedures (including contingency plan implementation) relevant to the positions in which they are employed.

(3) At a minimum, the training program must be designed to ensure that facility personnel are able to respond effectively to emergencies by familiarizing them with emergency procedures, emergency equipment, and emergency systems, including where applicable:

(i) Procedures for using, inspecting, repairing, and replacing facility emergency and monitoring equipment;

(ii) Key parameters for automatic waste feed cut-off systems;

(iii) Communications or alarm systems;

(iv) Response to fires or explosions;

(v) Response to ground-water contamination incidents; and

(vi) Shutdown of operations.

(b) Facility personnel must successfuly complete the program required in paragraph (a) of this section within six months after the effective date of these regulations or six months after the date of their employment or assignment to a facility, or to a new position at a facility, whichever is later. Employees hired after the effective date of these regulations must not work in unsupervised positions until they have completed the training requirements of paragraph (a) of this section.

(c) Facility personnel must take part in an annual review of the initial training required in paragraph (a) of this section.

(d) The owner or operator must maintain the following documents and records at the facility:

(1) The job title for each position at the facility related to hazardous waste management, and the name of the employee filling each job;

(2) A written job description for each position listed under paragraph (d)(1) of this Section. This description may be consistent in its degree of specificity with descriptions for other similar positions in the same company location or bargaining unit, but must include the requisite skill, education, or other qualifications, and duties of facility personnel assigned to each position;

(3) A written description of the type and amount of both introductory and continuing training that will be given to each person filling a position listed under paragraph (d)(1) of this section;

(4) Records that document that the training or job experience required under paragraphs (a), (b), and (c) of this section has been given to, and completed by, facility personnel.

(e) Training records on current personnel must be kept until closure of the facility. Training records on former employees must be kept for at least three years from the date the employee last worked at the facility. Personnel training racords may accompany personnel transferred within the same company.

§ 265.17 General requirements for ignitable, reactive, or incompatible wastes.

(a) The owner or operator must take precautions to prevent accidental ignition or reaction of ignitable or reactive waste. This waste must be separated and protected from sources of ignition or reaction including but not limited to: open flames, smoking, cutting and welding, hot surfaces, frictional heat, sparks (static, electrical, or mechanical), spontaneous ignition (e.g., from heat-producing chemical reactions), and radiant heat. While ignitable or reactive waste is being handled, the owner or operator must confine smoking and open flame to specially designated locations. "No Smoking" signs must be conspicuously placed wherever there is a hazard from ignitable or reactive waste.

(b) Where specifically required by other sections of this part, the treatment, storage, or disposal of ignitable or reactive waste, and the mixture or commingling of incompatible wastes, or incompatible wastes and materials, must be conducted so that it does not:

(1) Generate extreme heat or pressure, fire or explosion, or violent reaction;

(2) Produce uncontrolled toxic mists, fumes, dusts, or gases in sufficient quantities to threaten human health;

(3) Produce uncontrolled flammable fumes or gases in sufficient quantities to pose a risk of fire or explosions;

(4) Damage the structural integrity of the device or facility containing the waste; or

(5) Through other like means threaten human health or the environment.

Subpart C—Preparedness and Prevention

§ 265.30 Applicability.

The regulations in this subpart apply to owners and operators of all hazardous waste facilities, except as § 265.1 provides otherwise.

§ 265.31 Maintenance and operation of facility.

Facilities must be maintained and operated to minimize the possibility of a fire, explosion, or any unplanned sudden or non-sudden release of hazardous waste or hazardous waste constituents to air, soil, or surface water which could threaten human health or the environment.

§ 265.32 Required equipment.

All facilities must be equipped with the following, *unless* none of the hazards posed by waste handled at the facility could require a particular kind of equipment specified below:

(a) An internal communications or alarm system capable of providing immediate emergency instruction (voice or signal) to facility personnel;

(b) A device, such as a telephone (immediately available at the scene of operations) or a hand-held two-way radio, capable of summoning emergency assistance from local police departments, fire departments, or State or local emergency response teams;

(c) Portable fire extinguishers, fire control equipment (including special extinguishing equipment, such as that using foam, inert gas, or dry chemicals), spill control equipment, and decontamination equipment; and

(d) Water at adequate volume and pressure to supply water hose streams, or foam producing equipment, or automatic sprinklers, or water spray systems.

§ 265.33 Testing and maintenance of equipment.

All facility communications or alarm systems, fire protection equipment, spill control equipment, and decontamination equipment, where required, must be tested and maintained as necessary to assure its proper operation in time of emergency.

§ 265.34 Access to communications or alarm system.

(a) Whenever hazardous waste is being poured, mixed, spread, or otherwise handled, all personnel involved in the operation must have immediate access to an internal alarm or emergency communication device, either directly or through visual or voice contact with another employee, *unless* such a device is not required under § 265.32.

(b) If there is ever just one employee on the premises while the facility is operating, he must have immediate access to a device, such as a telephone (immediately available at the scene of operation) or a hand-held two-way radio, capable of summoning external emergency assistance, *unless* such a device is not required under § 265.32.

§ 265.35 Required aisle space.

The owner or operator must maintain aisle space to allow the unobstructed movement of personnel, fire protection equipment, spill control equipment, and decontamination equipment to any area of facility operation in an emergency, *unless* aisle space is not needed for any of these purposes.

§ 265.36 [Reserved]

§ 265.37 Arrangements with local authorities.

(a) The owner or operator must attempt to make the following arrangements, as appropriate for the type of waste handled at his facility and the potential need for the services of these organizations:

(1) Arrangements to familiarize police, fire departments, and emergency response teams with the layout of the facility, properties of hazardous waste handled at the facility and associated hazards, places where facility personnel would normally be working, entrances to roads inside the facility, and possible evacuation routes;

(2) Where more than one police and fire department might respond to an emergency, agreements designating primary emergency authority to a specific police and a specific fire department, and agreements with any others to provide support to the primary emergency authority;

(3) Agreements with State emergency response teams, emergency response contractors, and equipment suppliers; and

(4) Arrangements to familiarize local hospitals with the properties of hazardous waste handled at the facility and the types of injuries or illnesses which could result from fires, explosions, or releases at the facility.

(b) Where State or local authorities decline to enter into such arrangements, the owner or operator must document the refusal in the operating record.

Subpart D—Contingency Plan and Emergency Procedures

§ 265.50 Applicability.

The regulations in this subpart apply to owners and operators of all hazardous waste facilities, except as § 265.1 provides otherwise.

§ 265.51 Purpose and implementation of contingency plan.

(a) Each owner or operator must have a contingency plan for his facility. The contingency plan must be designed to minimize hazards to human health or the environment from fires, explosions, or any unplanned sudden or non-sudden release of hazardous waste or hazardous waste constituents to air, soil, or surface water.

(b) The provisions of the plan must be carried out immediately whenever there is a fire, explosion, or release of hazardous waste or hazardous waste constituents which could threaten human health or the environment.

§ 265.52   Content of contingency plan.

(a) The contingency plan must describe the actions facility personnel must take to comply with §§ 265.51 and 265.56 in response to fires, explosions, or any unplanned sudden or non-sudden release of hazardous waste or hazardous waste constituents to air, soil, or surface water at the facility.

(b) If the owner or operator has already prepared a Spill Prevention, Control, and Countermeasures (SPCC) Plan in accordance with Part 112 of this Chapter, or Part 1510 of Chapter V, or some other emergency or contingency plan, he need only amend that plan to incorporate hazardous waste management provisions that are sufficient to comply with the requirements of this Part.

(c) The plan must describe arrangements agreed to by local police departments, fire departments, hospitals, contractors, and State and local emergency response teams to coordinate emergency services, pursuant to § 265.37.

(d) The plan must list names, addresses, and phone numbers (office and home) of all persons qualified to act as emergency coordinator (see § 265.55), and this list must be kept up to date. Where more than one person is listed, one must be named as primary emergency coordinator and others must be listed in the order in which they will assume responsibility as alternates.

(e) The plan must include a list of all emergency equipment at the facility (such as fire extinguishing systems, spill control equipment, communications and alarm systems (internal and external), and decontamination equipment), where this equipment is required. This list must be kept up to date. In addition, the plan must include the location and a physical description of each item on the list, and a brief outline of its capabilities.

(f) The plan must include an evacuation plan for facility personnel where there is a possibility that evacuation could be necessary. This plan must describe signal(s) to be used to begin evacuation, evacuation routes, and alternate evacuation routes (in cases where the primary routes could be blocked by releases of hazardous waste or fires).

[45 FR 33233, May 19, 1980, as amended at 46 FR 27480, May 20, 1981]

§ 265.53   Copies of contingency plan.

A copy of the contingency plan and all revisions to the plan must be:

(a) Maintained at the facility; and

(b) Submitted to all local police departments, fire departments, hospitals, and State and local emergency response teams that may be called upon to provide emergency services.

§ 265.54   Amendment of contingency plan.

The contingency plan must be reviewed, and immediately amended, if necessary, whenever:

(a) Applicable regulations are revised;

(b) The plan fails in an emergency;

(c) The facility changes—in its design, construction, operation, maintenance, or other circumstances—in a way that materially increases the potential for fires, explosions, or releases of hazardous waste or hazardous waste constituents, or changes the response necessary in an emergency;

(d) The list of emergency coordinators changes; or

(e) The list of emergency equipment changes.

§ 265.55   Emergency coordinator.

At all times, there must be at least one employee either on the facility premises or on call (i.e., available to respond to an emergency by reaching the facility within a short period of time) with the responsibility for coordinating all emergency response measures. This emergency coordinator must be thoroughly familiar with all aspects of the facility's contingency plan, all operations and activities at the facility, the location and characteristics of waste handled, the location of all records within the facility, and the facility layout. In addition, this person must have the authority to commit the resources needed to carry out the contingency plan.

[*Comment:* The emergency coordinator's responsibilities are more fully spelled out in § 265.56. Applicable responsibilities for the emergency coordinator vary, depending on factors such as type and variety of waste(s) handled by the facility, and type and complexity of the facility.]

§ 265.56 Emergency procedures.

(a) Whenever there is an imminent or actual emergency situation, the emergency coordinator (or his designee when the emergency coordinator is on call) must immediately:

(1) Activate internal facility alarms or communication systems, where applicable, to notify all facility personnel; and

(2) Notify appropriate State or local agencies with designated response roles if their help is needed.

(b) Whenever there is a release, fire, or explosion, the emergency coordinator must immediately identify the character, exact source, amount, and a real extent of any released materials. He may do this by observation or review of facility records or manifests and, if necessary, by chemical analysis.

(c) Concurrently, the emergency coordinator must assess possible hazards to human health or the environment that may result from the release, fire, or explosion. This assessment must consider both direct and indirect effects of the release, fire, or explosion (e.g., the effects of any toxic, irritating, or asphyxiating gases that are generated, or the effects of any hazardous surface water run-offs from water or chemical agents used to control fire and heat-induced explosions).

(d) If the emergency coordinator determines that the facility has had a release, fire, or explosion which could threaten human health, or the environment, outside the facility, he must report his findings as follows:

(1) If his assessment indicates that evacuation of local areas may be advisable, he must immediately notify appropriate local authorities. He must be available to help appropriate officials decide whether local areas should be evacuated; and

(2) He must immediately notify either the government official designated as the on-scene coordinator for that geographical area (in the applicable regional contingency plan under Part 1510 of this Title), or the National Response Center (using their 24-hour toll free number 800/424-8802). The report must include:

(i) Name and telephone number of reporter;

(ii) Name and address of facility;

(iii) Time and type of incident (e.g., release, fire);

(iv) Name and quantity of material(s) involved, to the extent known;

(v) The extent of injuries, if any; and

(vi) The possible hazards to human health, or the environment, outside the facility.

(e) During an emergency, the emergency coordinator must take all reasonable measures necessary to ensure that fires, explosions, and releases do not occur, recur, or spread to other hazardous waste at the facility. These measures must include, where applicable, stopping processes and operations, collecting and containing released waste, and removing or isolating containers.

(f) If the facility stops operations in response to a fire, explosion or release, the emergency coordinator must monitor for leaks, pressure buildup, gas generation, or ruptures in valves, pipes, or other equipment, wherever this is appropriate.

(g) Immediately after an emergency, the emergency coordinator must provide for treating, storing, or disposing of recovered waste, contaminated soil or surface water, or any other material that results from a release, fire, or explosion at the facility.

[*Comment:* Unless the owner or operator can demonstrate, in accordance with § 261.3(c) or (d) of this Chapter, that the recovered material is not a hazardous waste, the owner or operator becomes a generator of hazardous waste and must manage it in accordance with all applicable requirements of Parts 262, 263, and 265 of this Chapter.]

(h) The emergency coordinator must ensure that, in the affected area(s) of the facility:

(1) No waste that may be incompatible with the released material is treated, stored, or disposed of until cleanup procedures are completed; and

(2) All emergency equipment listed in the contingency plan is cleaned and fit for its intended use before operations are resumed.

(i) The owner or operator must notify the Regional Administrator, and appropriate State and local authorities, that the facility is in compliance with paragraph (h) of this section before operations are resumed in the affected area(s) of the facility.

(j) The owner or operator must note in the operating record the time, date, and details of any incident that requires implementing the contingency plan. Within 15 days after the incident, he must submit a written report on the incident to the Regional Administrator. The report must include:

(1) Name, address, and telephone number of the owner or operator;

(2) Name, address, and telephone number of the facility;

(3) Date, time, and type of incident (e.g., fire, explosion);

(4) Name and quantity of material(s) involved;

(5) The extent of injuries, if any;

(6) An assessment of actual or potential hazards to human health or the environment, where this is applicable; and

(7) Estimated quantity and disposition of recovered material that resulted from the incident.

Subpart E—Manifest System, Recordkeeping, and Reporting

§ 265.70 Applicability.

The regulations in this subpart apply to owners and operators of both on-site and off-site facilities, except as § 265.1 provides otherwise. Sections 265.71, 265.72, and 265.76 do not apply to owners and operators of on-site facilities that do not receive any hazardous waste from off-site sources.

§ 265.71 Use of manifest system.

(a) If a facility receives hazardous waste accompanied by a manifest, the owner or operator, or his agent, must:

(1) Sign and date each copy of the manifest to certify that the hazardous waste covered by the manifest was received;

(2) Note any significant discrepancies in the manifest (as defined in § 265.72(a)) on each copy of the manifest;

[*Comment*: The Agency does not intend that the owner or operator of a facility whose procedures under § 265.13(c) include waste analysis must perform that analysis before signing the manifest and giving it to the transporter. Section 265.72(b), however, requires reporting an unreconciled discrepancy discovered during later analysis.]

(3) Immediately give the transporter at least one copy of the signed manifest;

(4) Within 30 days after the delivery, send a copy of the manifest to the generator; and

(5) Retain at the facility a copy of each manifest for at least three years from the date of delivery.

(b) If a facility receives, from a rail or water (bulk shipment) transporter, hazardous waste which is accompanied by a shipping paper containing all the information required on the manifest (excluding the EPA identification numbers, generator's certification, and signatures), the owner or operator, or his agent, must:

(1) Sign and date each copy of the manifest or shipping paper (if the manifest has not been received) to certify that the hazardous waste covered by the manifest or shipping paper was received;

(2) Note any significant discrepancies (as defined in § 265.72(a)) in the manifest or shipping paper (if the manifest has not been received) on each copy of the manifest or shipping paper;

[*Comment*: The Agency does not intend that the owner or operator of a facility whose procedures under § 265.13(c) include waste analysis must perform that analysis before signing the shipping paper and giving it to the transporter. Section 265.72(b), however, requires reporting an unreconciled discrepancy discovered during later analysis.]

(3) Immediately give the rail or water (bulk shipment) transporter at least one copy of the manifest or shipping paper (if the manifest has not been received);

(4) Within 30 days after the delivery, send a copy of the signed and dated manifest to the generator; however, if the manifest has not been received within 30 days after delivery, the owner or operator, or his agent, must send a copy of the shipping paper signed and dated to the generator; and

[*Comment:* Section 262.23(c) of this Chapter requires the generator to send three copies of the manifest to the facility when hazardous waste is sent by rail or water (bulk shipment).]

(5) Retain at the facility a copy of the manifest and shipping paper (if signed in lieu of the manifest at the time of delivery) for at least three years from the date of delivery.

(c) Whenever a shipment of hazardous waste is initiated from a facility, the owner or operator of that facility must comply with the requirements of Part 262 of this chapter.

[*Comment:* The provisions of § 262.34 are applicable to the on-site accumulation of hazardous wastes by generators. Therefore, the provisions of § 262.34 only apply to owners or operators who are shipping hazardous waste which they generated at that facility.]

[45 FR 33232, May 19, 1980, as amended at 45 FR 86970, 86974, Dec. 31, 1980]

§ 265.72   Manifest discrepancies.

(a) Manifest discrepancies are differences between the quantity or type of hazardous waste designated on the manifest or shipping paper, and the quantity or type of hazardous waste a facility actually receives. Significant discrepancies in quantity are: (1) for bulk waste, variations greater than 10 percent in weight, and (2) for batch waste, any variation in piece count, such as a discrepancy of one drum in a truckload. Significant discrepancies in type are obvious differences which can be discovered by inspection or waste analysis, such as waste solvent substituted for waste acid, or toxic constituents not reported on the manifest or shipping paper.

(b) Upon discovering a significant discrepancy, the owner or operator must attempt to reconcile the discrepancy with the waste generator or transporter (e.g., with telephone conversations). If the discrepancy is not resolved within 15 days after receiving the waste, the owner or operator must immediately submit to the Regional Administrator a letter describing the discrepancy and attempts to reconcile it, and a copy of the manifest or shipping paper at issue.

§ 265.73   Operating record.

(a) The owner or operator must keep a written operating record at his facility.

(b) The following information must be recorded, as it becomes available, and maintained in the operating record until closure of the facility:

(1) A description and the quantity of each hazardous waste received, and the method(s) and date(s) of its treatment, storage, or disposal at the facility as required by Appendix I;

(2) The location of each hazardous waste within the facility and the quantity at each location. For disposal facilities, the location and quantity of each hazardous waste must be recorded on a map or diagram of each cell or disposal area. For all facilities, this information must include cross-references to specific manifest document numbers, if the waste was accompanied by a manifest;

[*Comment:* See §§ 265.119, 265.279, and 265.309 for related requirements.]

(3) Records and results of waste analysis and trial tests performed as specified in §§ 265.13, 265.193, 265.225, 265.252, 265.273, 265.341, 265.375, and 265.402;

(4) Summary reports and details of all incidents that require implementing the contingency plan as specified in § 265.56(j);

(5) Records and results of inspections as required by § 265.15(d) (except these data need be kept only three years);

(6) Monitoring, testing, or analytical data where required by §§ 265.90, 265.94, 265.276, 265.278, 265.280(d)(1), 265.347, and 265.377; and,

[*Comment:* As required by § 265.94, monitoring data at disposal facilities must be kept throughout the post-closure period.]

(7) All closure cost estimates under § 265.142 and, for disposal facilities, all post-closure cost estimates under § 265.144.

[45 FR 33232, May 19, 1980, as amended at 46 FR 7680, Jan. 23, 1981]

§ 265.74 **Availability, retention, and disposition of records.**

(a) All records, including plans, required under this part must be furnished upon request, and made available at all reasonable times for inspection, by any officer, employee, or representative of EPA who is duly designated by the Administrator.

(b) The retention period for all records required under this Part is extended automatically during the course of any unresolved enforcement action regarding the facility or as requested by the Administrator.

(c) A copy of records of waste disposal locations and quantities under § 265.73(b)(2) must be submitted to the Regional Administrator and local land authority upon closure of the facility (see § 265.119).

§ 265.75 **Annual report.**

The owner or operator must prepare and submit a single copy of an annual report to the Regional Administrator by March 1 of each year. The report form and instructions in Appendix II must be used for this report. The annual report must cover facility activities during the previous calendar year and must include the following information:

(a) The EPA identification number, name, and address of the facility;

(b) The calendar year covered by the report;

(c) For off-site facilities, the EPA identification number of each hazardous waste generator from which the facility received a hazardous waste during the year; for imported shipments, the report must give the name and address of the foreign generator;

(d) A description and the quantity of each hazardous waste the facility received during the year. For off-site facilities, this information must be listed by EPA identification number of each generator;

(e) The method of treatment, storage, or disposal for each hazardous waste;

(f) Monitoring data under § 265.94(a)(2)(ii) and (iii), and (b)(2), where required;

(g) The most recent closure cost estimate under § 265.142, and, for disposal facilities, the most recent post-closure cost estimate under § 265.144; and

(h) The certification signed by the owner or operator of the facility or his authorized representative.

NOTE: At 47 FR 7842, Feb. 23, 1982, the compliance date for submission of generator and TSD facility 1981 annual reports, required under § 265.75 was extended to August 1, 1982.

§ 265.76 **Unmanifested waste report.**

If a facility accepts for treatment, storage, or disposal any hazardous waste from an off-site source without an accompanying manifest, or without an accompanying shipping paper as described in § 263.20(e)(2) of this Chapter, and if the waste is not excluded from the manifest requirement by § 261.5 of this Chapter, then the owner or operator must prepare and submit a single copy of a report to the Regional Administrator within 15 days after receiving the waste. The report form and instructions in Appendix II must be used for this report. The report must include the following information:

(a) The EPA identification number, name, and address of the facility;

(b) The date the facility received the waste;

(c) The EPA identification number, name, and address of the generator and the transporter, if available;

(d) A description and the quantity of each unmanifested hazardous waste the facility received;

(e) The method of treatment, storage, or disposal for each hazardous waste;

(f) The certification signed by the owner or operator of the facility or his authorized representative; and

(g) A brief explanation of why the waste was unmanifested, if known.

[Comment: Small quantities of hazardous waste are excluded from regulation under this Part and do not require a manifest.

Where a facility receives unmanifested hazardous wastes, the Agency suggests that the owner or operator obtain from each generator a certification that the waste qualifies for exclusion. Otherwise, the Agency suggests that the owner or operator file an unmanifested waste report for the hazardous waste movement.]

§ 265.77 **Additional reports.**

In addition to submitting the annual report and unmanifested waste reports described in §§ 265.75 and 265.76, the owner or operator must also report to the Regional Administrator:

(a) Releases, fires, and explosions as specified in § 265.56(j);

(b) Ground-water contamination and monitoring data as specified in §§ 265.93 and 265.94; and

(c) Facility closure as specified in § 265.115.

Subpart I—Use and Management of Containers

§ 265.170 **Applicability.**

The regulations in this Subpart apply to owners and operators of all hazardous waste facilities that store containers of hazardous waste, except as § 265.1 provides otherwise.

§ 265.171 **Condition of containers.**

If a container holding hazardous waste is not in good condition, or if it begins to leak, the owner or operator must transfer the hazardous waste from this container to a container that is in good condition, or manage the waste in some other way that complies with the requirements of this Part.

§ 265.172 Compatibility of waste with container.

The owner or operator must use a container made of or lined with materials which will not react with, and are otherwise compatible with, the hazardous waste to be stored, so that the ability of the container to contain the waste is not impaired.

§ 265.173 Management of containers.

(a) A container holding hazardous waste must always be closed during storage, except when it is necessary to add or remove waste.

(b) A container holding hazardous waste must not be opened, handled, or stored in a manner which may rupture the container or cause it to leak.

[Comment: Re-use of containers in transportation is governed by U.S. Department of Transportation regulations, including those set forth in 49 CFR 173.28.]

[45 FR 33232, May 19, 1980, as amended at 45 FR 78529, Nov. 25, 1980]

§ 265.174 Inspections.

The owner or operator must inspect areas where containers are stored, at least weekly, looking for leaks and for deterioration caused by corrosion or other factors.

[Comment: See § 265.171 for remedial action required if deterioration or leaks are detected.]

§ 265.175 [Reserved]

§ 265.176 Special requirements for ignitable or reactive waste.

Containers holding ignitable or reactive waste must be located at least 15 meters (50 feet) from the facility's property line.

[Comment: See § 265.17(a) for additional requirements.]

§ 265.177 Special requirements for incompatible wastes.

(a) Incompatible wastes, or incompatible wastes and materials, (see Appendix V for examples) must not be placed in the same container, unless § 265.17(b) is complied with.

(b) Hazardous waste must not be placed in an unwashed container that previously held an incompatible waste or material (see Appendix V for examples), unless § 265.17(b) is complied with.

(c) A storage container holding a hazardous waste that is incompatible with any waste or other materials stored nearby in other containers, piles, open tanks, or surface impoundments must be separated from the other materials or protected from them by means of a dike, berm, wall, or other device.

[Comment: The purpose of this is to prevent fires, explosions, gaseous emissions, leaching, or other discharge of hazardous waste or hazardous waste constituents which could result from the mixing of incompatible wastes or materials if containers break or leak.]

Subpart J—Tanks

§ 265.190 Applicability.

The regulations in this subpart apply to owners and operators of facilities that use tanks to treat or store hazardous waste, except as § 265.1 provides otherwise.

§ 265.191 [Reserved]

§ 265.192 General operating requirements.

(a) Treatment or storage of hazardous waste in tanks must comply with § 265.17(b).

(b) Hazardous wastes or treatment reagents must not be placed in a tank if they could cause the tank or its inner liner to rupture, leak, corrode, or otherwise fail before the end of its intended life.

(c) Uncovered tanks must be operated to ensure at least 60 centimeters (2 feet) of freeboard, unless the tank is equipped with a containment structure (e.g., dike or trench), a drainage control system, or a diversion structure (e.g., standby tank) with a capacity that equals or exceeds the volume of the top 60 centimeters (2 feet) of the tank.

(d) Where hazardous waste is continuously fed into a tank, the tank must be equipped with a means to stop this inflow (e.g., a waste feed cutoff system or by-pass system to a stand-by tank).

[*Comment:* These systems are intended to be used in the event of a leak or overflow from the tank due to a system failure (e.g., a malfunction in the treatment process, a crack in the tank, etc.).]

§ 265.193   Waste analysis and trial tests.

(a) In addition to the waste analysis required by § 265.13, whenever a tank is to be used to:

(1) Chemically treat or store a hazardous waste which is substantially different from waste previously treated or stored in that tank; or

(2) Chemically treat hazardous waste with a substantially different process than any previously used in that tank; the owner or operator must, before treating or storing the different waste or using the different process:

(i) Conduct waste analyses and trial treatment or storage tests (e.g., bench scale or pilot plant scale tests); or

(ii) Obtain written, documented information on similar storage or treatment of similar waste under similar operating conditions;

to show that this proposed treatment or storage will meet all applicable requirements of § 265.192(a) and (b).

[*Comment:* As required by § 265.13, the waste analysis plan must include analyses needed to comply with §§ 265.198 and 265.199. As required by § 265.73, the owner or operator must place the results from each waste analysis and trial test, or the documented information, in the operating record of the facility.]

§ 265.194   Inspections.

(a) The owner or operator of a tank must inspect, where present:

(1) Discharge control equipment (e.g., waste feed cut-off systems, by-pass systems, and drainage systems), at least once each operating day, to ensure that it is in good working order;

(2) Data gathered from monitoring equipment (e.g., pressure and temperature gauges), at least once each operating day, to ensure that the tank is being operated according to its design;

(3) The level of waste in the tank, at least once each operating day, to ensure compliance with § 265.192(c);

(4) The construction materials of the tank, at least weekly, to detect corrosion or leaking of fixtures or seams; and

(5) The construction materials of, and the area immediately surrounding, discharge confinement structures (e.g., dikes), at least weekly, to detect erosion or obvious signs of leakage (e.g., wet spots or dead vegetation).

[*Comment:* As required by § 265.15(c), the owner or operator must remedy any deterioration or malfunction he finds.]

§§ 265.195—265.196   [Reserved]

§ 265.197   Closure.

At closure, all hazardous waste and hazardous waste residues must be removed from tanks, discharge control equipment, and discharge confinement structures.

[*Comment:* At closure, as throughout the operating period, unless the owner or operator can demonstrate, in accordance with § 261.3(c) or (d) of this Chapter, that any solid waste removed from his tank is not a hazardous waste, the owner or operator becomes a generator of hazardous waste and must manage it in accordance with all applicable requirements of Parts 262, 263, and 265 of this Chapter.]

§ 265.198   Special requirements for ignitable or reactive waste.

(a) Ignitable or reactive waste must not be placed in a tank, unless:

(1) The waste is treated, rendered, or mixed before or immediately after placement in the tank so that (i) the resulting waste, mixture, or dissolution of material no longer meets the definition of ignitable or reactive waste under §§ 261.21 or 261.23 of this Chapter, and (ii) § 265.17(b) is complied with; or

(2) The waste is stored or treated in such a way that it is protected from any material or conditions which may cause the waste to ignite or react; or

(3) The tank is used solely for emergencies.

(b) The owner or operator of a facility which treats or stores ignitable or reactive waste in covered tanks must comply with the buffer zone requirements for tanks contained in Tables 2-1 through 2-6 of the National Fire Protection Association's "Flammable and Combustible Liquids Code" (1977 or 1981), (incorporated by reference, see § 260.11).

[45 FR 33232, May 19, 1980, as amended at 46 FR 35249, July 7, 1981]

§ 265.199 Special requirements for incompatible wastes.

(a) Incompatible wastes, or incompatible wastes and materials, (see Appendix V for examples) must not be placed in the same tank, unless § 265.17(b) is complied with.

(b) Hazardous waste must not be placed in an unwashed tank which previously held an incompatible waste or material, unless § 265.17(b) is complied with.

Subpart K—Surface Impoundments

§ 265.220 Applicability.

The regulations in this Subpart apply to owners and operators of facilities that use surface impoundments to treat, store, or dispose of hazardous waste, except as § 265.1 provides otherwise.

§ 265.221 [Reserved]

§ 265.222 General operating requirements.

A surface impoundment must maintain enough freeboard to prevent any overtopping of the dike by overfilling, wave action, or a storm. There must be at least 60 centimeters (2 feet) of freeboard.

[*Comment:* Any point source discharge from a surface impoundment to waters of the United States is subject to the requirements of Section 402 of the Clean Water Act, as amended. Spills may be subject to Section 311 of that Act.]

§ 265.223 Containment system.

All earthen dikes must have a protective cover, such as grass, shale, or rock, to minimize wind and water erosion and to preserve their structural integrity.

§ 265.224 [Reserved]

§ 265.225 Waste analysis and trial tests.

(a) In addition to the waste analyses required by § 265.13, whenever a surface impoundment is to be used to:

(1) Chemically treat a hazardous waste which is substantially different from waste previously treated in that impoundment; or

(2) Chemically treat hazardous waste with a substantially different process than any previously used in that impoundment; the owner or operator must, before treating the different waste or using the different process:

(i) Conduct waste analyses and trial treatment tests (e.g., bench scale or pilot plant scale tests); or

(ii) Obtain written, documented information on similar treatment of similar waste under similar operating conditions; to show that this treatment will comply with § 265.17(b).

[*Comment:* As required by § 265.13, the waste analysis plan must include analyses needed to comply with §§ 265.229 and 265.230. As required by § 265.73, the owner or operator must place the results from each waste analysis and trial test, or the documented information, in the operating record of the facility.]

§ 265.226 Inspections.

(a) The owner or operator must inspect:

(1) The freeboard level at least once each operating day to ensure compliance with § 265.222, and

(2) The surface impoundment, including dikes and vegetation surrounding the dike, at least once a week to detect any leaks, deterioration, or failures in the impoundment.

[*Comment:* As required by § 265.15(c), the owner or operator must remedy any deterioration or malfunction he finds.]

§ 265.227 [Reserved]

§ 265.228 Closure and post-closure.

(a) At closure, the owner or operator may elect to remove from the impoundment:

(1) Standing liquids;

(2) Waste and waste residues;

(3) The liner, if any; and

(4) Underlying and surrounding contaminated soil.

(b) If the owner or operator removes all the impoundment materials in paragraph (a) of this section, or can demonstrate under § 261.3(c) and (d) of this chapter that none of the materials listed in paragraph (a) of this Section remaining at any stage of removal are hazardous wastes, the impoundment is not further subject to the requirements of this Part.

[*Comment:* At closure, as throughout the operating period, unless the owner or operator can demonstrate, in accordance with § 261.3 (c) or (d) of this chapter, that any solid waste removed from the surface impoundment is not a hazardous waste, he becomes a generator of hazardous waste and must manage it in accordance with all applicable requirements of Parts 262, 263, and 265 of this chapter. The surface impoundment may be subject to Part 257 of this Chapter even if it is not subject to this Part.]

(c) If the owner or operator does not remove all the impoundment materials in paragraph (a) of this section, or does not make the demonstration in paragraph (b) of this section, he must close the impoundment and provide post-closure care as for a landfill under Subpart G and § 265.310. If necessary to support the final cover specified in the approved closure plan, the owner or operator must treat remaining liquids, residues, and soils by removal of liquids, drying, or other means.

[*Comment:* The closure requirements under § 265.310 will vary with the amount and nature of the residue remaining, if any, and the degree of contamination of the underlying and surrounding soil. Section 265.117(d) allows the Regional Administrator to vary post-closure care requirements.]

§ 265.229 Special requirements for ignitable or reactive waste.

(a) Ignitable or reactive waste must not be placed in a surface impoundment, unless:

(1) The waste is treated, rendered, or mixed before or immediately after placement in the impoundment so that (i) the resulting waste, mixture, or dissolution of material no longer meets the definition of ignitable or reactive waste under §§ 261.21 or 261.23 of this Chapter, and (ii) § 265.17(b) is complied with; or

(2) The surface impoundment is used solely for emergencies.

§ 265.230 Special requirements for incompatible wastes.

Incompatible wastes, or incompatible wastes and materials, (see Appendix V for examples) must not be placed in the same surface impoundment, unless § 265.17(b) is complied with.

Subpart L—Waste Piles

§ 265.250 Applicability.

The regulations in this subpart apply to owners and operators of facilities that treat or store hazardous waste in piles, except as § 265.1 provides otherwise. Alternatively, a pile of hazardous waste may be managed as a landfill under Subpart N.

§ 265.251 Protection from wind.

The owner or operator of a pile containing hazardous waste which could be subject to dispersal by wind must cover or otherwise manage the pile so that wind dispersal is controlled.

§ 265.252 Waste analysis.

In addition to the waste analyses required by § 265.13, the owner or operator must analyze a representative sample of waste from each incoming movement before adding the waste to any existing pile, *unless* (1) The only wastes the facility receives which are amenable to piling are compatible with each other, or (2) the waste received is compatible with the waste in the pile to which it is to be added. The analysis conducted must be capable of differentiating between the types of hazardous waste the owner or operator places in piles, so that mixing of incompatible waste does not inadvertently occur. The analysis must include a visual comparison of color and texture.

[*Comment:* As required by § 265.13, the waste analysis plan must include analyses needed to comply with §§ 265.256 and 265.257. As required by § 265.73, the owner or operator must place the results of this analysis in the operating record of the facility.]

§ 265.253 Containment.

If leachate or run-off from a pile is a hazardous waste, then either:

(a) The pile must be placed on an impermeable base that is compatible with the waste under the conditions of treatment or storage, run-on must be diverted away from the pile, and any leachate and run-off from the pile must be collected and managed as a hazardous waste; or

(b)(1) The pile must be protected from precipitation and run-on by some other means; and

(2) No liquids or wastes containing free liquids may be placed in the pile.

[*Comment:* If collected leachate or run-off is discharged through a point source to waters of the United States, it is subject to the requirements of Section 402 of the Clean Water Act, as amended.]

(c) The date for compliance with paragraphs (a) and (b)(1) of this section is 12 months after the effective date of this Part.

§§ 265.254—265.255 [Reserved]

§ 265.256 Special requirements for ignitable or reactive waste.

(a) Ignitable or reactive wastes must not be placed in a pile, *unless:*

(1) Addition of the waste to an existing pile (i) results in the waste or mixture no longer meeting the definition of ignitable or reactive waste under §§ 261.21 or 261.23 of this Chapter, and (ii) complies with § 265.17(b); or

(2) The waste is managed in such a way that it is protected from any material or conditions which may cause it to ignite or react.

§ 265.257 Special requirements for incompatible wastes.

(a) Incompatible wastes, or incompatible wastes and materials, (see Appendix V for examples) must not be placed in the same pile, unless § 265.17(b) is complied with.

(b) A pile of hazardous waste that is incompatible with any waste or other material stored nearby in other containers, piles, open tanks, or surface impoundments must be separated from the other materials, or protected from them by means of a dike, berm, wall, or other device.

[*Comment:* The purpose of this is to prevent fires, explosions, gaseous emissions, leaching, or other discharge of hazardous waste or hazardous waste constituents which could result from the contact or mixing of incompatible wastes or materials.]

(c) Hazardous waste must not be piled on the same area where incompatible wastes or materials were previously piled, unless that area has been decontaminated sufficiently to ensure compliance with § 265.17(b).

Regulations on Transportation of Hazardous Chemicals

DESCRIPTION OF WASTE CATEGORIES

DOT has classified materials for transport purposes according to the type of hazard they present. These hazard classes are listed in Table D.1. The definitions of each class, with references to the corresponding section of 49 CFR, are abstracted in Table D.2. DOT requires that materials that pose more than one of these hazards be classified according to the highest hazard class, in the hierarchy shown in Table D.3 (49 CFR 173.2).

TABLE D.1 DOT Hazard Classes

Combustible liquid[a]	Flammable liquid	ORM-B
Corrosive material[b]	Flammable solid	ORM-E
Etiologic agent	Irritating materials	Oxidizer
Class A explosive	Nonflammable gas	Poison A
Class B explosive	Organic peroxide	Poison B
Class C explosive	ORM-A	Radioactive material
Flammable gas		

[a]There is no corresponding class of combustible solid; see Table D.2.
[b]Although there is only one corrosive hazard class, the hazard relative to a Poison B depends on whether a corrosive material is liquid or solid; see Table D.3.

219

TABLE D.2 Abstracts of Definitions of DOT Hazard Classes (The numbers that follow each heading refer to sections of 49 CFR.)

Combustible, 173.115(2)(b)(1).
A liquid having a flash point at or above 37.8°C (100°F) and below 98.3°C (200°F). A solid that melts below 60°C (140°F) and that has a flash point in the above temperature range when liquid can be listed as a combustible liquid.

Corrosive Material, 173.240(a) and (b).
A liquid or solid that causes visible destruction or irreversible alterations in human skin tissue at the site of contact, or in the case of leakage from its packaging, a liquid that has a severe corrosion rate on steel.
    1. A material is considered to be destructive or to cause irreversible alteration in human skin tissue if, when tested on the intact skin of the albino rabbit by the technique described in Appendix A of Part 173, the structure of the tissue at the site of contact is destroyed or changed irreversibly after an exposure period of 4 hours or less.
    2. A liquid is considered to have a severe corrosion rate if its corrosion rate exceeds 6.35 mm/year (0.25 in./year) on SAE 1020 steel at a test temperature of 55°C (130°F). An acceptable test is described in National Association of Corrosion Engineers Standard TM-01-69.
    3. If human experience indicates that the hazard of the material is greater than would be indicated by test results of 1 or 2.

Etiologic Agent, 173.386.
A viable microorganism, or its toxin, which causes or may cause human disease.

Explosive, 173.50.
An explosive is any chemical compound, mixture, or device the primary or common purpose of which is to function by explosion, i.e., with substantially instantaneous release of gas or heat, unless such compound, mixture, or device is otherwise specifically classified in Parts 170–189 of 49 CFR.
    Forbidden explosives, which are not permitted to be transported, include compounds or mixtures that ignite spontaneously or undergo marked decomposition when subjected to a temperature of 75°C (167°F) for 48 consecutive hours (173.51).

Class A Explosive, 173.53.
Detonating or other types of maximum hazard. There are nine types of Class A explosives. Those that might be encountered in laboratories other than those working on explosive materials or devices are solids or liquids that can be detonated when unconfined by initiation, e.g., by a safety fuse, electrical squib, or impact in a prescribed test. Examples include trinitrotoluene, picric acid, urea nitrate, heavy-metal azides and fulminates, and nitroglycerine.

Class B Explosive, 173.88.
Explosives that function by rapid combustion rather than detonation. Generally includes explosive devices such as special fireworks or flash powder. Pose a flammable hazard.

Class C Explosive, 173.100.
Certain types of manufactured articles containing Class A or Class B explosives, or both, as components but in restricted quantities, and certain types of fireworks. Minimum hazard.

TABLE D.2 (*continued*)

Flammable Gas, 173.300(b).
A compressed gas that meets the characteristics of lower flammability limit, flammability limit range, flame projection, or flame propagation as specified in 173.300(b).

Flammable Liquid, 173.115(a)(1).
A liquid having a flash point below 37.8°C (100°F).

Flammable Solid, 173.150.
A solid, other than an explosive, that is liable to cause fires through friction, absorption of moisture, spontaneous chemical change, retained heat from processing, or that can be ignited readily and when ignited burns so vigorously and persistently as to create a serious transportation hazard. This class includes spontaneously combustible and water-reactive materials.

Irritating Material, 173.381.
A liquid or solid substance that on contact with fire or when exposed to air gives off dangerous or intensely irritating fumes, which cause reversible local irritant effects on eyes, nose, and throat, temporarily impairing a person's ability to function, such as bromobenzyl cyanide, chloroacetophenone, diphenylaminochloroarsine, and diphenylchloroarsine, but not including any poisonous material, Class A.

Nonflammable Gas, 173.300(a).
Any material or mixture having in the container a pressure exceeding 40 psi(abs) at 21°C (70°F) or having an absolute pressure exceeding 104 psi(abs) at 54.4°C (130°F) or any flammable liquid having a vapor pressure exceeding 40 psi(abs) at 37.8°C (100°F).

Organic Peroxide, 173.151a.
An organic compound containing the bivalent -O-O- structure and that can be considered a derivative of hydrogen peroxide where one or more of the hydrogen atoms have been replaced by organic radicals, unless . . . [see 173.151(a) for exceptions].

ORM (Other Regulated Material), 173.500.
Other Regulated Material (ORM) is a material that does not meet the definition of a hazardous material, other than a combustible liquid in packagings having a capacity of 416 L (110 gallons) or less and is specified in Table 172.101 as an ORM material or that possesses one or more of the characteristics described in the following groups.

ORM-A, 173.500(b)(1).
A material that has an anesthetic, irritating, noxious, toxic, or other similar property and that can cause extreme annoyance or discomfort to passengers and crew in the event of leakage during transportation.

ORM-B, 173.500(b)(2).
A material (including a solid wet with water) capable of causing significant damage to a transport vehicle from leakage during transportation. Materials meeting one or both of the following criteria are ORM-B materials:

TABLE D.2   *(continued)*

1. A liquid that has a corrosion rate exceeding 6.35 mm/year (0.25 in./year) on aluminum (nonclad 7075-T6) at 54.5°C (130°F).
2. Specifically designated by name in Table 172.101.

ORM-E, 173.500(b)(5).
A material that is not included in any other hazard class, but includes
1. Hazardous waste.
2. Hazardous substances described in 49 CFR 171.8.

Oxidizer, 173.151.
A substance, such as a chlorate, permanganate, inorganic peroxide, or a nitrate, that yields oxygen readily to stimulate the combustion of organic matter.

Poison A, 173.326.
Extremely dangerous poisons. Poisonous gases or liquids of such nature that a very small amount of the gas, or vapor of the liquid, mixed with air is dangerous to life. This class includes the following:
Bromoacetone
Cyanogen
Cyanogen chloride containing less than 0.9% water
Diphosgene
Ethyldichloroarsine
Hydrocyanic acid
Methyldichloroarsine
Nitrogen tetroxide
Nitrogen tetroxide–nitric oxide mixture containing up to 33.2% by weight nitric oxide
Phosgene

Poison B, 173.343.
Less dangerous than Poisons A. Substances, liquids, or solids, other than Class A poisons or Irritating Materials, that are known to be so toxic to man as to afford a hazard to health during transportation or that, in the absence of adequate data on human toxicity, are presumed to be toxic to man. This includes the following:

1. A chemical that has an LD_{50} of 50 mg/kg or less when administered orally to albino rats weighing 200–300 g each.
2. A chemical that has an LD_{50} of 200 mg/kg or less when administered by continuous contact for 24 h or less with the bare skin of albino rabbits weighing 2–3 kg each.
3. A chemical that has an LC_{50} in air of 200 ppm or less when administered by continuous inhalation for 1 h or less to albino rats weighing 200–300 g each, provided such concentration is likely to be encountered by man when the chemical is used in any reasonably foreseeable manner.
4. If available data on human experience indicate results different from those obtained on animals, the human data shall take precedence.

Radioactive Material, 173.389.
A material or combination of materials that spontaneously emits ionizing radiation, and having a specific activity greater than 0.002 µCi per gram.

TABLE D.3 Hierarchy of Decreasing Hazard of DOT Classes[a]

DOT Class	
Radioactive	Irritating agent
Poison A	Combustible liquid (in containers with a
Flammable gas	capacity greater than 416 L (110 gal-
Nonflammable gas	lons)
Flammable liquid	ORM-B
Oxidizer	ORM-A
Flammable solid	Combustible liquid (in containers with a
Corrosive (liquid)	capacity of 416 L (110 gallons) or less
Poison B	ORM-E
Corrosive (solid)	

[a]This hierarchy does not apply to materials listed in Table 172.101 or to explosives, etiologic agents, or organic peroxides.

DOT has listed the appropriate hazard class of many commercial materials in Table 172.101 of 49 CFR 172.101. The table also includes a variety of information necessary for the description, packaging, and labeling of those materials. This table is essential for generators who package and ship their own waste.

PACKAGING, LABELING, AND MARKING OF WASTE

Hazardous waste must be properly packaged in accordance with the general shipping and packaging requirements in 49 CFR 173. The most direct route to locating the appropriate sections in Part 173 is through columns 5a and 5b in Table 172.101. Section 173 paragraphs in turn refer to shipping container specifications in 49 CFR 178.

The common outside containers for shipment are 55-gallon closed-head (DOT 17C) or open-head (DOT 17H) steel drums. These are single-trip containers (STC) and are stamped with that identification and the drum style. Smaller containers can be used if they meet DOT specifications for materials.

Some shippers prefer to use new drums or drums reconditioned to DOT standards to minimize the chance of leakage during handling and transport. During shipment, drums are subjected to significant physical stress through bouncing and vibration and must therefore be free of creases and significant corrosion.

Unreconditioned STC drums previously used to transport a hazardous material can be used to transport an authorized hazardous waste if the following conditions are met [49 CFR 173.28(p)]:

(a) Packaging must comply with DOT regulations on the reuse of containers (49 CFR 173.28).
(b) The packages can be transported only by highway.
(c) Packages must not be offered for transportation less than 24 h after closure, and each package must be inspected for leakage immediately prior to being offered for transportation.
(d) The packages must be loaded by the shipper and unloaded by the consignee, unless the carrier is a private or contract carrier.

Each container of hazardous waste must be marked and labeled in accordance with DOT regulations (49 CFR 172 Subparts D and E). Effective 1 July 1983, DOT requires that containers of a hazardous substance as defined by DOT (49 CFR 171.8) be marked to include the name of the substance and the letters RQ as part of the proper shipping name. Proper shipping name markings on these packages may include the word "Waste," and its inclusion is recommended.

EPA regulations further require that containers of hazardous waste be marked with the following (40 CFR 262.32):

(a) Generator's name and address
(b) Manifest document number
(c) The following statement: "HAZARDOUS WASTE—Federal law prohibits improper disposal. If found, contact the nearest police or public safety authority or the U.S. Environmental Protection Agency."

States may require additional markings such as the registration number assigned by the state to the waste category and an identification number assigned by the coordinator for the individual container. Where interstate shipments are involved, such information may be required by the states in which the generator and the disposal facilities are located. The generator may also elect to include other information such as the 12-digit identification number assigned to the generator by EPA.

Containers must be labeled with the appropriate DOT hazard labels indicated in column 4 of Table 172.101. Label designs and specifications are specified in 49 CFR 172.400–.446. Labels are available from commercial label distributors. Additional labels may be required when the generator has information that a material meets the definition of more than one hazard class (49 CFR 172.402). When hazardous materials that have different hazard classes are packed within the same outside container, such as a lab pack, the container must be labeled as required for each class of hazardous material (49 CFR 172.404). Samples of materials listed in Table 172.101 may often be exempted from labeling and specification packaging requirements when they are being surface transported for analysis (see guide to exemption packaging in Table 172.101, column 5a). Poison A and B materials are special in that exception packaging may be allowed but exception labeling is not allowed (49 CFR 173.345 and 49 CFR 173.364). If a reasonable doubt exists as to the hazard class and labeling requirements for a sample, labeling can be based on the shipper's judgment and the hierarchy of hazard classes in Table D.2 [49 CFR 172.402(h)].

Markings on containers must be able to withstand exposure to the sun and water. A generator of large quantities of waste may find it useful to develop a stencil system where markings are made by a fast-drying spray paint.

PREPARATION OF SHIPPING PAPERS AND HAZARDOUS-WASTE MANIFEST

1. DOT SHIPPING PAPERS

DOT requires that shipping papers in the form of a shipping order, bill of lading, manifest, or other shipping document serving the same purpose accompany the shipment if the material is a hazardous waste (49 CFR Subpart C). The following information is required:

Name of shipper
Shipping description (DOT proper shipping name, hazard class, and UN/NA identification number as required by 49 CFR 172.101, 172.202, and 172.203)
Total quantity of material
Certification of proper classification, description, packaging, marking, labeling, and proper condition
Signature of a representative of the generator

DOT definitions for each hazard class are summarized in Table D.2. Note that the hazard class ORM-E is for materials whose hazard is not described by one of the higher-priority hazard classes.

When a material only presents one hazard and the proper shipping name clearly indicates the hazard class, the hazard class need not be included as a separate part of the shipping description. When the waste in a container has more than one DOT hazard, each hazard should be shown in the proper shipping name. The hazards should be listed in the priority sequence shown in Table D.3. In this case the priority hazard class should also be entered between the proper shipping name and the identification number.

A precise name should be used when the contents of a container are homogeneous and are listed by name in the table. More typically, laboratory waste will be a mixture, and a generic description should be used.

The generator must insert the word "waste" before the proper shipping name shown in the table.

Commonly used generic proper shipping descriptions from Table 172.101 for waste mixtures include the following[a]:

Waste Corrosive Liquid, n.o.s.	UN1760
Waste Corrosive Liquid, Poisonous, n.o.s.	UN2922
Waste Corrosive Solid, n.o.s.	UN1759
Waste Flammable Liquid, Corrosive, n.o.s.	UN2924
Waste Flammable Liquid, n.o.s.	UN1993
Waste Flammable Liquid, Poisonous, n.o.s.	UN1992
Waste Flammable Solid, Corrosive, n.o.s.	UN2925
Waste Flammable Solid, n.o.s.	UN1325
Waste Flammable Solid, Poisonous, n.o.s.	UN2926
Hazardous Waste, Liquid, n.o.s., ORM-E	NA9189
Hazardous Waste, Solid, n.o.s., ORM-E	NA9189
Waste Poisonous Liquid, Flammable, n.o.s.	NA1953
Waste Poisonous Liquid, n.o.s.	NA1955
Waste Poison B, Liquid, n.o.s.	UN2810
Waste Poison B, Solid, n.o.s.	UN2811
Waste Poisonous Solid, Corrosive, n.o.s.	UN2928

[a]n.o.s., not otherwise specified.

Chemicals that are classed as DOT Poison A and Poison B must be listed by chemical name in the shipping description (49 CFR 172.203k) on the shipping paper. However, an exemption that eliminates the need for listing more than one poison in a lab pack can be requested from the DOT Office of Hazardous Materials Regulation (49 CFR 107.111).

If a material qualifies as a DOT-designated hazardous substance and the quantity in any container exceeds the minimum reportable quantity (RQ) designated in Table 172.101, column 2, the letters "RQ" must also be entered in the basic description.

Uniform straight bill of lading forms can generally be obtained from stationery or printing firms. A sample bill of lading is shown in Figure D.1.

2. HAZARDOUS-WASTE MANIFEST (MANIFEST)

EPA and DOT require that a manifest signed by representatives of the generator and the carrier accompany any shipment (40 CFR Part 262; 49 CFR Section 172.205).

STRAIGHT BILL OF LADING — SHORT FORM — ORIGINAL — NOT NEGOTIABLE

RECEIVED, subject to the classifications and tariffs in effect on the date of the issue of this Bill of Lading.

From   XYZ Chemical Company

SHIPPER'S NO	DATE SHIPPED	CAR OR VEHICLE INITIALS & NUMBER
	2/28/83	123-XYZ

ROUTE

Anytown/Boonville/Baton Rouge

DELIVERY ADDRESS (TO BE FILLED IN ONLY WHEN SHIPPER DESIRES AND GOVERNING TARIFFS PROVIDE FOR DELIVERY THEREAT)

12345 Industrial Boulevard, Baton Rouge, LA

CONSIGNED TO   (MAIL OR STREET ADDRESS OF CONSIGNEE — FOR PURPOSES OF NOTIFICATION ONLY)

Waste Hauling Inc., 123 State St., Boonville, MO

DESTINATION (CITY, STATE, AND COUNTY)

Baton Rouge, LA

No Packages	HM	Kind of Package, Description of Articles, Special Marks, and Exceptions	*Weight (Sub. to Cor.)	Class or rate	Ck Col	
21	X	Waste Flammable Liquid NOS	10221			Subject to Section 7 of Conditions of applicable bill of lading, if this shipment is to be delivered to the consignee without recourse on the consignor, the consignor shall sign the following statement. The carrier shall not make delivery of this shipment without payment of freight and all other lawful charges.
		UN 1993				
		Drum NOS. 1-21, incl.				
						(Signature of consignor)
						If charges are to be prepaid, write or stamp here, "To be Prepaid"

*If the shipment moves between two ports by a carrier by water, the law requires that the bill of lading shall state whether it is carrier's or shippers weight. Where the rate is dependent on value, the agreed or declared value of the property is hereby specifically stated by the shipper to be not exceeding $1.50 per pound

This is to certify that the above named materials are properly classified, described, packaged, marked, and labeled, and are in proper condition for transportation, according to the applicable regulations of the Department of Transportation.

The fibre boxes used for this shipment conform to the specifications set forth in the box maker's certificate thereon, and all other requirements of Consolidated Freight Classification

(Shipper's imprint in lieu of stamp, not a part of bill of lading approved by the Interstate Commerce Commission.)

SHIPPER, PER John P. Smith   Carrier's Agent Joe Driver Per _____

FOR CHEMICAL EMERGENCY SPILL, LEAK, FIRE, EXPOSURE OR ACCIDENT CALL CHEMTREC — DAY OR NIGHT 1-800-424-9300

BL IR 5 REV 3 82

FIGURE D.1

Sufficient copies must be made to provide one each for the generator, each transporter, and the operator of the disposal facility and one to be returned to the generator when completely signed.

The manifest must specify one disposal facility that has an EPA identification number and that has a permit to operate under existing state or local regulations. The manifest must contain the following information:

Manifest document number. The manifest document number must be a serially increasing number assigned by the generator. Each page should have a distinct document number if more than one page is required for the necessary information.

Generator's name, mailing address, telephone number, and EPA identification number.

Name and EPA identification number of each transporter.

228

Form DNR H.W.G.-10
HAZARDOUS WASTE MANIFEST DOCUMENT

MANIFEST DOCUMENT NUMBER

1	2	3	4	5		0	0	2		0	0	2
					Generator I.D. No.				Waste I.D. No.		Shipment No.	

Part 1 to be completed by the generator   (Instructions for completing and handling this document are on the reverse side)

Name

	Identification	Address		Telephone No.	Date Shipped or Rec'd.

Item 1. Generator
XYZ Chemical Company — Generator I.D. No. 12345 — 2121 Anystreet Anytown, MO — (314) 321-25XX — 2/28/83

Item 2. Transporter
Waste Hauling, Inc. — Transporter No. GDT-898 — 123 State Street Boonville, MO — (314) 987-65YY — 2/28/83

Item 3. Treatment, Storage or Disposal Facility
Waste Landfill Company — T,S,D, Facility Permit No. LAD-007-215-767 — P.O. Box 100 New Orleans, LA — (504) 999-88ZZ — 3/2/83

Item 4.
Proper DOT Shipping Name: Hazardous Waste Solid NOS NA 9189 — DOT Hazard Class ORM-E — DOT Label Required or Exceptions ORM-E — Quantity 22 — Units* 1 2 3 G 5 — Weight (if applicable) 7221

*Circle one: 1. tons; 2. gallons; 3. cubic yds; 4. drums - 55 gallon; or 5. Pounds

Item 5. Immediate Emergency Response Information
In the event of a spill, contact the National Response Center, U.S. Coast Guard, 800-424-8802
SPECIAL HANDLING INSTRUCTIONS

24-hour emergency telephone numbers
(314) 321-25AA
Chemtrec 800-424-9300

Item 6. Placards Provided or Affixed — None

Shipper's Check List

X DOT Labels Applied and Secure	X DOT Auth. Containers
X Proper DOT Name on all Packages	X Checked for Proper Sealing
Air Cargo Only	Peligro Label Applied

Item 7. GENERATOR CERTIFICATION. This is to certify that the above named materials are properly classified, described, packaged, marked, and labeled, and are in proper condition for transportation according to the applicable regulations of the Department of Transportation and the _____ Department of Natural Resources.

Generator's Signature John P. Smith Date 2/28/83

Part 2
To be completed by the transporter
Item 8. TRANSPORTER CERTIFICATION. This is to certify acceptance of the hazardous waste shipment. Date accepted for Shipment:
Transporter's Signature Joe Driver Date 2/28/83
Part 3
Item 9. TSDF CERTIFICATION. This is to certify acceptance of the hazardous waste for treatment, storage or disposal.
TSDF Signature Fred Jones Date 3/2/83

Department Final Copy

FIGURE D.2

Name, address, and EPA identification number of the designated treatment, storage, or disposal facility and alternative facility, if any.

Description of the hazardous waste.

The total quantity of each hazardous waste by units of weight or volume and the type and number of containers as loaded in or on the transport vehicle.

The following certification must appear on the manifest: "This is to certify that the above named materials are properly classified, described, packaged, marked and labeled and are in proper condition for transportation according to the applicable regulations of the Department of Transportation and the EPA." The above certification must be signed by hand by the generator.

If the manifest contains all the information required by, and is prepared in accordance with, 49 CFR Subpart C as incidated above, it can be used to satisfy the DOT hazardous-material shipping paper requirement [49 CFR 172.205(h)]. States from and to which hazardous waste is transported may require use of their state hazardous-waste manifest in addition to that required by DOT. A separate state manifest may also be required for intrastate shipments. Sample state manifests are shown in Figures D.2 and D.3.

LOADING FOR SHIPMENT

The DOT regulations include specific instructions for the loading and unloading of hazardous waste (49 CFR Subpart B). These include requirements that the handbrake on the vehicle be set and that smoking be forbidden.

Trucks should be loaded using a hand truck or forklifts as appropriate to the facility. The loading sequence may be determined by the driver. The generator should be satisfied that the load is secured against movement on the truck.

Before leaving the premises, the driver must complete the transporter's section of the manifest. The generator must then obtain and retain a copy with the signature of the driver. The remaining copies are carried by the transporter in the cab of the truck.

RECORDKEEPING AND REPORTING

An Exception Report must be filed by the generator with the EPA Regional Administrator if a copy of the manifest is not returned within 45 days of the shipping date. States may also require reporting to state

Form DNR H.W.G. - 10
HAZARDOUS WASTE MANIFEST DOCUMENT

MANIFEST DOCUMENT NUMBER

1	2	3	4	5	0	0	1	0	0	1
Generator I.D. No.					Waste I.D. No.			Shipment No.		

Part 1 to be completed by the generator   (Instructions for completing and handling this document are on the reverse side)

Name	Identification	Address	Telephone No.	Date Shipped or Rec'd.
Item 1. Generator XYZ Chemical Company	Generator I.D. No. 12345	2121 Anystreet Anytown, MO	(314) 321-25XX	2/28/83
Item 2. Transporter Waste Hauling, Inc.	Transporter No. GDT-898	123 State Street, Boonville, MO	(314) 987-65YY	2/28/83
Item 3. Treatment, Storage or Disposal Facility Waste Incineration, Inc.	T,S,D, Facility Permit No. LAD-006-212-989	P.O. Box 100 Baton Rouge, LA	(504) 999-8822	3/2/83

Item 4. Proper DOT Shipping Name	DOT Hazard Class	DOT Label Required or Exceptions	Quantity	Units*	Weight (if applicable)
Waste Flammable Liquid NOS UN 1993	Flammable	Flammable	21	1 2 3 0 5	10221

*Circle one: 1. tons; 2. gallon; 3. cubic yds; 4. drums - 55 gallon; or 5. Pounds

Item 5.	Immediate Emergency Response Information		Item 6.	Placards Provided or Affixed
	In the event of a spill, contact the National Response Center, U.S. Coast Guard, 800-424-8802	24-hour emergency telephone numbers (314) 321-25AA Chemtrec 800-424-9300		Flammable

Flammable

Shipper's Check List

X	DOT Labels Applied and Secure	X	DOT Auth. Containers
X	Proper DOT Name on all Packages	X	Checked for Proper Sealing
	Air Cargo Only		Peligro Label Applied

SPECIAL HANDLING INSTRUCTIONS

Item 7. GENERATOR CERTIFICATION. This is to certify that the above named materials are properly classified, described, packaged, marked, and labeled, and are in proper condition for transportation according to the applicable regulations of the Department of Transportation and the   Department of Natural Resources.

Generator's Signature ___John P. Smith___   Date ___2/28/83___

Part 2

To be completed by the transporter

Item 8. TRANSPORTER CERTIFICATION. This is to certify acceptance of the hazardous waste shipment. Date accepted for Shipment:

Transporter's Signature ___Joe Driver___   Date ___2/28/83___

Part 3

Item 9. TSDF CERTIFICATION. This is to certify acceptance of the hazardous waste for treatment, storage or disposal.

TSDF Signature ___Bob Burner___   Date ___3/2/83___

Department Final Copy

FIGURE D.3

agencies. The Exception Report consists of a copy of the manifest and a cover letter signed by the generator explaining the efforts taken to locate the hazardous waste and the results of those efforts (40 CFR 262.42).

The generator is required to keep for 3 years a copy of each signed manifest from the disposal facility that received the hazardous waste. The generator is also required to keep a copy of each Exception Report for a period for 3 years from the date of the report (40 CFR 262.40).

A generator must keep records of any test results, waste analysis, or other determination made in accordance with Section 40 CFR 262.11 for at least 3 years from the date that the waste was last sent to on-site or off-site storage or disposal. The periods for retention cited above are automatically extended during the course of any unresolved enforcement action regarding the regulated activity or as requested by the EPA Administrator. Generators that use an internal manifest system to track waste from a laboratory to a packaging area for shipment may want to retain any records that serve to document the composition of the waste.

EPA may require that generators file an Annual Report covering waste handling activities. States may have separate reporting requirements, typically quarterly or annually. States may also require that a copy of the completed manifest covering waste generated in, or shipped to, their state be filed with the appropriate agency.

E Incompatible Chemicals

The term "incompatible chemicals" refers to chemicals that can react with each other

- Violently
- With evolution of substantial heat
- To produce flammable products or
- To produce toxic products

The EPA RCRA regulations specify that incompatible chemicals must not be placed in the same lab pack for landfill disposal, and the DOT regulations have a similar proscription on packing incompatible chemicals for transport. Incompatible chemicals should always be handled, stored, and packed so that they cannot accidently come into contact with each other.

Guidelines for the segregation of common laboratory chemicals that are incompatible are presented in Tables E.1 and E.2. Table E.1 contains general classes of compounds that should be kept separated; Table E.2 lists specific compounds that can pose reactivity hazards. Chemicals in each grouping in columns A and B of each table should be kept separate. Additional information on specific chemical reaction hazards can be found in the the following references:

Manual of Hazardous Chemical Reactions, A Compilation of Chemical Reactions Reported to be Potentially Hazardous, National Fire Protection Association, NFPA 491M, 1975, NFPA, 470 Atlantic Avenue, Boston, Mass. 02210.

L. Bretherick, *Handbook of Reactive Chemical Hazards*, 2nd ed., Butterworths, London-Boston, 1979.

L. Bretherick, ed., *Hazards in the Chemical Laboratory*, 3rd ed., Royal Society of Chemistry, London, 1981.

TABLE E.1   General Classes of Incompatible Chemicals

A	B
Acids	Bases, metals
Oxidizing agents[a]	*Reducing agents[a]*
Chlorates	Ammonia, anhydrous and
Chromates	aqueous
Chromium trioxide	Carbon
Dichromates	Metals
Halogens	Metal hydrides
Halogenating agents	Nitrites
Hydrogen peroxide	Organic compounds
Nitric acid	Phosphorus
Nitrates	Silicon
Perchlorates	Sulfur
Peroxides	
Permanganates	
Persulfates	

[a]The examples of oxidizing and reducing agents are illustrative of common laboratory chemicals; they are not intended to be exhaustive.

TABLE E.2   Specific Chemical Incompatibilities

A	B
Acetylene and monosubstituted acetylenes ($RC{\equiv}CH$)	Group IB and IIB metals and their salts Halogens Halogenating agents
Ammonia, anhydrous and aqueous	Halogens Halogenating agents Mercury Silver

Alkali and alkaline earth
  carbides
  hydrides
  hydroxides
  metals
  oxides
  peroxides
	Water
	Acids
	Halogenated organic compounds
	Halogenating agents
	Oxidizing agents[a]

Azides, inorganic	Acids Heavy metals and their salts Oxidizing agents[a]
Cyanides, inorganic	Acids Strong bases
Mercury and its amalgams	Acetylene Ammonia, anhydrous and aqueous Nitric acid Sodium azide
Nitrates, inorganic	Acids Reducing agents[a]
Nitric acid	Bases Chromic acid Chromates Metals Permanganates Reducing agents[a] Sulfides Sulfuric acid
Nitrites, inorganic	Acids Oxidizing agents[a]

TABLE E.2 (*continued*)

A	B
Organic compounds	Oxidizing agents[a]
Organic acyl halides	{ Bases
	Organic hydroxy and amino compounds
Organic anhydrides	{ Bases
	Organic hydroxy and amino compounds
Organic halogen compounds	{ Group IA and IIA metals
	Aluminum
Organic nitro compounds	Strong bases
Oxalic acid	{ Mercury and its salts
	Silver and its salts
Phosphorus	{ Oxidizing agents[a]
	Oxygen
	Strong bases
Phosphorus pentoxide	{ Alcohols
	Strong bases
	Water
Sulfides, inorganic	Acids
Sulfuric acid (concentrated)	{ Bases
	Potassium permanganate
	Water

[a]See list of examples in Table E.1.

APPENDIX

F

Potentially Explosive Chemicals and Reagent Combinations

Table F.1 lists some common classes of laboratory chemicals that have potential for producing a violent explosion when subjected to shock or friction. These chemicals should never be disposed of as such but should be handled by procedures suggested in Chapters 6 and 7. Information on these and some less common classes of explosives has been gathered by Bretherick.[1]

Table F.2 lists a few illustrative combinations of common laboratory reagents that can produce explosions when they are brought together or that give reaction products that can explode without any apparent external initiating action. This list is by no means exhaustive; additional information of potentially explosive reagent combinations can be found in Reference 2.

TABLE F.1   Shock-Sensitive Compounds

Acetylenic compounds, especially polyacetylenes, haloacetylenes, and heavy metal salts of acetylenes (copper, silver, and mercury salts are particularly sensitive)

Acyl nitrates

Alkyl nitrates, particularly polyol nitrates such as nitrocellulose and nitroglycerine

Alkyl and acyl nitrites

Alkyl perchlorates

Amminemetal oxosalts: metal compounds with coordinated ammonia, hydrazine, or similar nitrogenous donors and ionic perchlorate, nitrate, permanganate, or other oxidizing group

Azides, including metal, nonmetal, and organic azides

Chlorite salts of metals, such as $AgClO_2$ and $Hg(ClO_2)_2$

Diazo compounds such as CH_2N_2

Diazonium salts, when dry

Fulminates (silver fulminate, AgCNO, can form in the reaction mixture from the Tollens' test for aldehydes if it is allowed to stand for some time; this can be prevented by adding dilute nitric acid to the test mixture as soon as the test has been completed)

Hydrogen peroxide becomes increasingly treacherous as the concentration rises above 30%, forming explosive mixtures with organic materials and decomposing violently in the presence of traces of transition metals

N-Halogen compounds such as difluoroamino compounds and halogen azides

N-Nitro compounds such as N-nitromethylamine, nitrourea, nitroguanidine, and nitric amide

Oxo salts of nitrogenous bases: perchlorates, dichromates, nitrates, iodates, chlorites, chlorates, and permanganates of ammonia, amines, hydroxylamine, guanidine, etc.

Perchlorate salts. Most metal, nonmetal, and amine perchlorates can be detonated and may undergo violent reaction in contact with combustible materials

Peroxides and hydroperoxides, organic (see Chapter 6, Section II.P)

Peroxides (solid) that crystallize from or are left from evaporation of peroxidizable solvents (see Chapter 6 and Appendix I).

Peroxides, transition-metal salts

Picrates, especially salts of transition and heavy metals, such as Ni, Pb, Hg, Cu, and Zn; picric acid is explosive but is less sensitive to shock or friction than its metal salts and is relatively safe as a water-wet paste (see Chapter 7)

Polynitroalkyl compounds such as tetranitromethane and dinitroacetonitrile

Polynitroaromatic compounds, especially polynitro hydrocarbons, phenols, and amines

TABLE F.2 Potentially Explosive Combinations of Some Common Reagents

Acetone + chloroform in the presence of base
Acetylene + copper, silver, mercury, or their salts
Ammonia (including aqueous solutions) + Cl_2, Br_2, or I_2
Carbon disulfide + sodium azide
Chlorine + an alcohol
Chloroform or carbon tetrachloride + powdered Al or Mg
Decolorizing carbon + an oxidizing agent
Diethyl ether + chlorine (including a chlorine atmosphere)
Dimethyl sulfoxide + an acyl halide, $SOCl_2$, or $POCl_3$
Dimethyl sulfoxide + CrO_3
Ethanol + calcium hypochlorite
Ethanol + silver nitrate
Nitric acid + acetic anhydride or acetic acid
Picric acid + a heavy-metal salt, such as of Pb, Hg, or Ag
Silver oxide + ammonia + ethanol
Sodium + a chlorinated hydrocarbon
Sodium hypochlorite + an amine

REFERENCES

1. L. Bretherick, *Handbook of Reactive Chemical Hazards*, 2nd ed., Butterworths, London-Boston, 1979, p. 60.
2. *Manual of Hazardous Chemical Reactions, A Compilation of Chemical Reactions Reported to be Potentially Hazardous*, National Fire Protection Association, NFPA 491M, 1975, NFPA, 470 Atlantic Avenue, Boston, Mass. 02210.

Water-Reactive
G Chemicals

This appendix lists some common laboratory chemicals that react violently with water and that should always be stored and handled so that they do not come into contact with liquid water or water vapor. They are prohibited from landfill disposal, even in a lab pack, because of the characteristic of reactivity. Procedures for decomposing laboratory quantities are given in Chapter 6, with the pertinent section given in parentheses.

Alkali metals (III.D)
Alkali metal hydrides (III.C.2)
Alkali metal amides (III.C.7)
Metal alkyls, such as lithium alkyls and aluminum alkyls (IV.A)
Grignard reagents (IV.A)
Halides of nonmetals, such as BCl_3, BF_3, PCl_3, PCl_5, $SiCl_4$, S_2Cl_2 (III.F)
Inorganic acid halides, such as $POCl_3$, $SOCl_2$, SO_2Cl_2 (III.F)
Anhydrous metal halides, such as $AlCl_3$, $TiCl_4$, $ZrCl_4$, $SnCl_4$ (III.E)
Phosphorus pentoxide (III.I)
Calcium carbide (IV.E)
Organic acid halides and anhydrides of low molecular weight (II.J)

APPENDIX

H

Pyrophoric
Chemicals

Many members of the following readily oxidized classes of common laboratory chemicals ignite spontaneously in air. A more extensive list that includes less common chemicals can be found in Reference 1. Pyrophoric chemicals should be stored in tightly closed containers under an inert atmosphere (or, for some, an inert liquid), and all transfers and manipulations of them must be carried out under an inert atmosphere or liquid. Pyrophoric chemicals cannot be put into a landfill because of the characteristic of reactivity.[2] Suggested disposal procedures are in the sections of Chapter 6 given in parentheses after each class.

Grignard reagents, RMgX (IV.A)
Metal alkyls and aryls, such as RLi, RNa, R_3Al, R_2Zn (IV.A)
Metal carbonyls, such as $Ni(CO)_4$, $Fe(CO)_5$, $Co_2(CO)_8$ (IV.B)
Alkali metals such as Na, K (III.D.1)
Metal powders, such as Al, Co, Fe, Mg, Mn, Pd, Pt, Ti, Sn, Zn, Zr (III.D.2)
Metal hydrides, such as NaH, $LiAlH_4$ (IV.C.2)
Nonmetal hydrides, such as B_2H_6 and other boranes, PH_3, AsH_3 (III.G)
Nonmetal alkyls, such as R_3B, R_3P, R_3As (IV.C)
Phosphorus (white) (III.H)

240

REFERENCES

1. L. Bretherick, *Handbook of Reactive Chemical Hazards*, 2nd ed., Butterworths, London-Boston, 1979, pp. 167–171.
2. Appendix A [40 CFR 265.316(e)].

Peroxide-Forming
Chemicals

Many common laboratory chemicals can form peroxides when allowed access to air over a period of time. A single opening of a container to remove some of the contents can introduce enough air for peroxide formation to occur. Some types of compounds form peroxides that are treacherously and violently explosive in concentrated solution or as solids. Accordingly, peroxide-containing liquids should never be evaporated to dryness. Peroxide formation can also occur in many polymerizable unsaturated compounds, and these peroxides can initiate a runaway, sometimes explosive polymerization reaction. Procedures for testing for peroxides and for removing small amounts from laboratory chemicals are given in Chapter 6, Section II.P.

This appendix provides a list of structural characteristics in organic compounds that can peroxidize and some common inorganic materials that form peroxides. Although the tabulation of organic structures may seem to include a large fraction of the common organic chemicals, they are listed in an approximate order of decreasing hazard. Reports of serious incidents involving the last five organic structural types are extremely rare, but they are included because laboratory workers should be aware that they can form peroxides that can influence the course of experiments in which they are used.

This appendix also provides specific examples of common chemicals that can become serious hazards because of peroxide formation. Suggested time limits are given for retention or testing of these compounds after opening the original container. Although some laboratories mark containers of such chemicals with the date of receipt of the original container, it should be recognized that such dating does not take into account the unknown time span between original packaging of the chemical and its date of receipt. The date of opening the original container of a chemical that is a hazardous peroxide-former should always be marked on the container. Labels such as that illustrated below should be provided to all laboratory workers to affix to and date all samples of peroxide-forming reagents that they receive.

The material in this appendix has been adapted from Reference 1. This and Reference 2 should be consulted for additional information on labeling and handling of peroxide-forming chemicals.

Table I.2 gives examples of common laboratory chemicals that are prone to form peroxides on exposure to air. The lists are not exhaustive, and analogous compounds that have any of the structural features given in Table I.1 should be tested for the presence of peroxides before being used as solvents or being distilled. The recommended retention times begin with the date of synthesis or of opening the original container.

PEROXIDIZABLE COMPOUND

	RECEIVED	OPENED
DATE	_____	_____

DISCARD OR TEST WITHIN 6 MONTHS
AFTER OPENING

OR-10058

TABLE I.1   Types of Chemicals That Are Prone to Form Peroxides

A. Organic Structures (in approximate order of decreasing hazard)

1. $\overset{\displaystyle H}{\underset{\diagup}{\overset{\diagdown}{C}}}-O-$ Ethers and acetals with α hydrogen atoms

2. $\overset{\diagdown}{\underset{\diagup}{C}}=\overset{\displaystyle H}{\underset{}{C}}-\overset{\diagup}{\underset{\diagdown}{C}}$ Olefins with allylic hydrogen atoms

3. $\overset{\diagdown}{\underset{\diagup}{C}}=\overset{\displaystyle X}{\underset{}{C}}-$ Chloroolefins and fluoroolefins

4. $CH_2{=}C\overset{\diagup}{\diagdown}$ Vinyl halides, esters, and ethers

5. $\overset{\diagdown}{\underset{\diagup}{C}}=C-C=\overset{\diagup}{\underset{\diagdown}{C}}$ Dienes

6. $\overset{\diagdown}{\underset{\diagup}{C}}=\overset{\displaystyle H}{\underset{}{C}}-C{\equiv}CH$ Vinylacetylenes with α hydrogen atoms

7. $\overset{\displaystyle H}{\underset{\diagup}{\overset{\diagdown}{C}}}-C{\equiv}CH$ Alkylacetylenes with α hydrogen atoms

8. $\overset{\displaystyle H}{\underset{\diagup}{\overset{\diagdown}{C}}}-Ar$ Alkylarenes that contain tertiary hydrogen atoms

9. $-\overset{\displaystyle |}{\underset{|}{C}}-H$ Alkanes and cycloalkanes that contain tertiary hydrogen atoms

10. $\overset{\diagdown}{\underset{\diagup}{C}}=\overset{\displaystyle |}{\underset{}{C}}-CO_2R$ Acrylates and methacrylates

TABLE I.1   (*continued*)

11.	H \ | C—OH /	Secondary alcohols
12.	O H ‖ | / —C—C \	Ketones that contain α hydrogen atoms
13.	H | —C=O	Aldehydes
14.	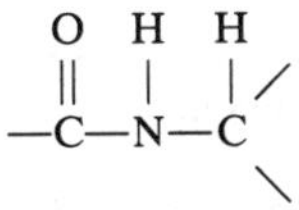 O H H ‖ | | / —C—N—C \	Ureas, amides, and lactams that have a hydrogen atom on a carbon atom attached to nitrogen.

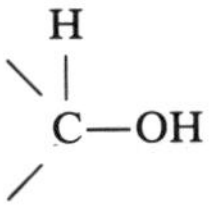

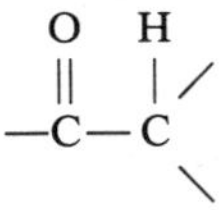

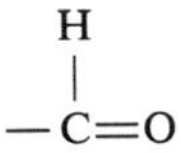

B. *Inorganic Substances*

    1. Alkali metals, especially potassium, rubidium, and cesium (see Chapter 6, Section III.D)

    2. Metal amides (see Chapter 6, Section III.C.7)

    3. Organometallic compounds with a metal atom bonded to carbon (see Chapter 6, Section IV)

    4. Metal alkoxides

TABLE I.2   Common Peroxide-Forming Chemicals

LIST A
Severe Peroxide Hazard on Storage with Exposure to Air
Discard within 3 months

- Diisopropyl ether (isopropyl ether)
- Divinylacetylene (DVA)[a]
- Potassium metal
- Potassium amide
- Sodium amide (sodamide)
- Vinylidene chloride (1,1-dichloro-ethylene)[a]

LIST B
Peroxide Hazard on Concentration; Do Not Distill or Evaporate Without First Testing for the Presence of Peroxides
Discard or test for peroxides after 6 months

- Acetaldehyde diethyl acetal (acetal)
- Cumene (isopropylbenzene)
- Cyclohexene
- Cyclopentene
- Decalin (decahydronaphthalene)
- Diacetylene (butadiene)
- Dicyclopentadiene
- Diethyl ether (ether)
- Diethylene glycol dimethyl ether (diglyme)

TABLE I.2   *(continued)*

• Dioxane	• Furan
• Ethylene glycol dimethyl ether (glyme)	• Methylacetylene
	• Methylcyclopentane
• Ethylene glycol ether acetates	• Methyl isobutyl ketone
• Ethylene glycol monoethers (cellosolves)	• Tetrahydrofuran (THF)
	• Tetralin (tetrahydronaphthalene)
	• Vinyl ethers[a]

LIST C

Hazard of Rapid Polymerization Initiated by Internally Formed Peroxides[a]

a. *Normal Liquids; Discard or test for peroxides after 6 months[b]*

• Chloroprene (2-chloro-1,3-butadiene)[c]	• Vinyl acetate
	• Vinylpyridine
• Styrene	

b. *Normal Gases; Discard after 12 months[d]*

• Butadiene[c]	• Vinylacetylene (MVA)[c]
• Tetrafluoroethylene (TFE)[c]	• Vinyl chloride

[a]Polymerizable monomers should be stored with a polymerization inhibitor from which the monomer can be separated by distillation just before use.

[b]Although common acrylic monomers such as acrylonitrile, acrylic acid, ethyl acrylate, and methyl methacrylate can form peroxides, they have not been reported to develop hazardous levels in normal use and storage.

[c]The hazard from peroxides in these compounds is substantially greater when they are stored in the liquid phase, and if so stored without an inhibitor they should be considered as in LIST A.

[d]Although air will not enter a gas cylinder in which gases are stored under pressure, these gases are sometimes transferred from the original cylinder to another in the laboratory, and it is difficult to be sure that there is no residual air in the receiving cylinder. An inhibitor should be put into any such secondary cylinder before one of these gases is transferred into it; the supplier can suggest inhibitors to be used. The hazard posed by these gases is much greater if there is a liquid phase in such a secondary container, and even inhibited gases that have been put into a secondary container under conditions that create a liquid phase should be discarded within 12 months.

REFERENCES

1. H. L. Jackson, W. B. McCormack, C. S. Rondesvedt, K. C. Smeltz, and I. E. Viele, *Safety in the Chemical Laboratory,* Vol. 3, N. V. Steere, ed., reprinted from *J. Chem. Educ.*, Division of Chemical Health and Safety, American Chemical Society, Easton, Pa. 18042.
2. NRC Committee on Hazardous Substances in the Laboratory, *Prudent Practices for Handling Hazardous Chemicals in the Laboratory*, National Academy Press, Washington, D.C., 1981, pp. 63–65.

pH Ranges for Precipitating Hydroxides of Cations[1]

Most metal ions are precipitated as hydroxides or oxides at high pH. However, many precipitates will redissolve in excess base. For this reason it is necessary to control pH closely in a number of cases. The following table shows the recommended pH range for precipitating many cations in their most common oxidation state. The notation "1 N" in the right-hand column indicates that the precipitate will not dissolve in 1 N sodium hydroxide (pH 14).

pH RANGE FOR PRECIPITATION OF METAL HYDROXIDES AND OXIDES

	1	2	3	4	5	6	7	8	9	10	
Ag^{1+}									├── ▶		1 N
Al^{3+}						├──────┤					
As^{3+}	Not Precipitated (precipitate as sulfide)										
As^{5+}	Not Precipitated (precipitate as sulfide)										
Au^{3+}							├──────┤				
Be^{2+}							├──────┤				
Bi^{3+}							├──────────────▶				1 N
Cd^{2+}							├──────────────▶				1 N
Co^{2+}								├────────▶			1 N
Cr^{3+}							├──────────────▶				1 N

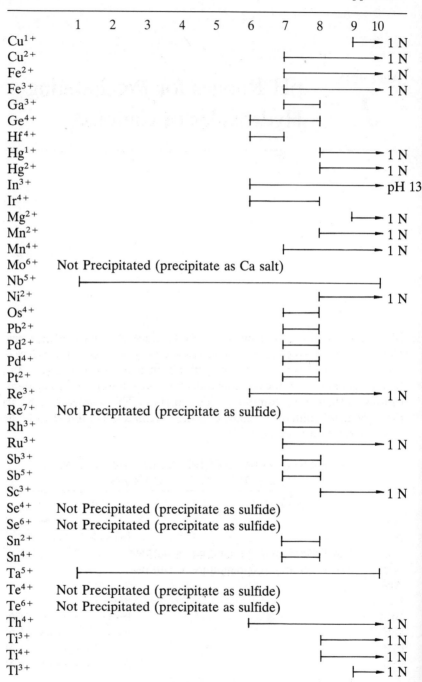

	1	2	3	4	5	6	7	8	9	10	
Cu^{1+}									⊢	→	1 N
Cu^{2+}							⊢			→	1 N
Fe^{2+}							⊢			→	1 N
Fe^{3+}							⊢			→	1 N
Ga^{3+}							⊢—⊣				
Ge^{4+}						⊢		⊣			
Hf^{4+}						⊢—⊣					
Hg^{1+}								⊢		→	1 N
Hg^{2+}							⊢			→	1 N
In^{3+}						⊢				→	pH 13
Ir^{4+}						⊢		⊣			
Mg^{2+}									⊢	→	1 N
Mn^{2+}								⊢		→	1 N
Mn^{4+}							⊢			→	1 N
Mo^{6+}	Not Precipitated (precipitate as Ca salt)										
Nb^{5+}	⊢										⊣
Ni^{2+}									⊢	→	1 N
Os^{4+}						⊢		⊣			
Pb^{2+}						⊢		⊣			
Pd^{2+}						⊢		⊣			
Pd^{4+}						⊢		⊣			
Pt^{2+}						⊢		⊣			
Re^{3+}						⊢				→	1 N
Re^{7+}	Not Precipitated (precipitate as sulfide)										
Rh^{3+}						⊢—⊣					
Ru^{3+}						⊢				→	1 N
Sb^{3+}						⊢		⊣			
Sb^{5+}						⊢		⊣			
Sc^{3+}									⊢	→	1 N
Se^{4+}	Not Precipitated (precipitate as sulfide)										
Se^{6+}	Not Precipitated (precipitate as sulfide)										
Sn^{2+}							⊢	⊣			
Sn^{4+}							⊢	⊣			
Ta^{5+}	⊢										⊣
Te^{4+}	Not Precipitated (precipitate as sulfide)										
Te^{6+}	Not Precipitated (precipitate as sulfide)										
Th^{4+}							⊢			→	1 N
Ti^{3+}								⊢		→	1 N
Ti^{4+}								⊢		→	1 N
Tl^{3+}									⊢	→	1 N

	1	2	3	4	5	6	7	8	9	10
V^{4+}							├──┤			
V^{5+}							├──┤			
W^{6+}	Not Precipitated (precipitate as Ca salt)									
Zn^{2+}								├──┤		
Zr^{4+}						├──┤				

REFERENCE

1. L. Erdey, *Gravimetric Analysis, Part II*, Pergamon Press, New York, 1965; D. T. Burns, A. Townsend, and A. H. Carter, *Inorganic Reaction Chemistry*, Vol. 2, Ellis Horwood, New York, 1981.

Guidelines for Disposal of Chemicals in the Sanitary Sewer System

The following lists comprise compounds that are suitable for disposal down the drain with excess water in quantities up to about 100 g at a time. However, local regulations may prohibit drain disposal of some and should be checked before any laboratory compiles its list of compounds acceptable for disposal down its drains. Compounds on both lists are water soluble to at least 3% and present low toxicity hazard. Those on the organic list are readily biodegradable.

I. ORGANIC CHEMICALS

ALCOHOLS

Alkanols with less than 5 carbon atoms
t-Amyl alcohol
Alkanediols with less than 8 carbon atoms
Glycerol
Sugars and sugar alcohols
Alkoxyalkanols with less than 7 carbon atoms
$n\text{-}C_4H_9OCH_2CH_2OCH_2CH_2OH$
2-Chloroethanol

ALDEHYDES

Aliphatic aldehydes with less than 5 carbon atoms

AMIDES

$RCONH_2$ and RCONHR with less than 5 carbon atoms
$RCONR_2$ with less than 11 carbon atoms

AMINES[a]

Aliphatic amines with less than 7 carbon atoms
Aliphatic diamines with less than 7 carbon atoms
Benzylamine
Pyridine

CARBOXYLIC ACIDS

Alkanoic acids with less than 6 carbon atoms[a]
Alkanedioic acids with less than 6 carbon atoms
Hydroxyalkanoic acids with less than 6 carbon atoms
Aminoalkanoic acids with less than 7 carbon atoms
Ammonium, sodium, and potassium salts of the above acid classes with
    less than 21 carbon atoms
Chloroalkanedioic acids with less than 4 carbon atoms

ESTERS

Esters with less than 5 carbon atoms
Isopropyl acetate

ETHERS

Tetrahydrofuran
Dioxolane
Dioxane

[a]Those with a disagreeable odor, such as dimethylamine, 1,4-butanediamine, butyric acids, and valeric acids, should be neutralized, and the resulting salt solutions flushed down the drain, diluted with at least 1000 volumes of water.

ment

KETONES

Ketones with less than 6 carbon atoms

NITRILES

Acetonitrile
Propionitrile

SULFONIC ACIDS

Sodium or potassium salts of most are acceptable

II. INORGANIC CHEMICALS

This list comprises water-soluble compounds of low-toxic-hazard cations and low-toxic-hazard anions. Compounds of any of these ions that are strongly acidic or basic should be neutralized before disposal down the drain.

Cations	*Anions*
Al^{3+}	BO_3^{3-}, $B_4O_7^{2-}$
Ca^{2+}	Br^-
Cu^{2+}	CO_3^{2-}
$Fe^{2+,3+}$	Cl^-
H^+	HSO_3^-
K^+	OCN^-
Li^+	OH^-
Mg^{2+}	I^-
Na^+	NO_3^-
NH_4^+	PO_4^{3-}
Sn^{2+}	SO_4^{2-}
Sr^{2+}	SCN^-
$Ti^{3+,4+}$	
Zn^{2+}	
Zr^{2+}	

APPENDIX
L

Incineration Equipment

I. INCINERATION SYSTEMS

A. SUSPENSION FIRING

In suspension firing, waste is injected into a hot combustion chamber (firebox) in such a way that the waste is suspended in the hot combustion gases until the waste is thoroughly oxidized. The ash portion of the waste is either entrained by the flue gases leaving the firebox (flyash) or falls to the bottom of the firebox (bottom ash).

An incinerator designed for injection of liquid wastes should have an atomizer capable of injecting a fine spray of liquid waste into the firebox to ensure complete oxidation of the liquid. Pressure or mechanical atomizers are cheap and reliable but require high liquid pressure (at least 300 psig to generate a fairly fine spray of droplets less than 500 μm in diameter). Very fine sprays are best produced by twin-fluid atomizers, which utilize the energy of a second fluid, such as compressed air or steam, to form small liquid droplets less than 100 μm in diameter.

Normally the liquid waste is fed from a storage tank to the incinerator under flow or pressure control. Auxiliary fuel is used to maintain the desired firebox temperature.

A portion of the process diagram for an incinerator for handling highly chlorinated liquid hydrocarbon wastes is shown in Figure L.1. The flue gas treatment techniques utilized in chlorinated hydrocarbon incinerators are discussed below in Section II.

Solids reduced to a size at which their burnout times are less than the

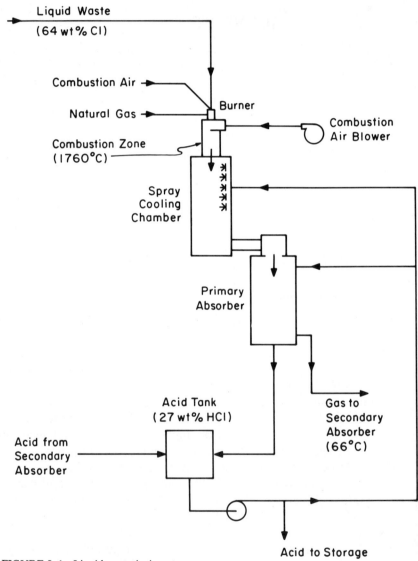

FIGURE L.1 Liquid-waste incinerator.

residence time of the gases in a combustion chamber (about 2 seconds) can be burned in suspension. The pulverized solids are transported pneumatically with part of the combustion air. They burn in a two-stage process, with part of them volatilizing and burning in the gas phase and the remainder burning heterogeneously. The volatile fraction usually stabilizes the flame. Experience on combustion of solids has been gained with pulverized coal, usually milled to a weight average diameter of 50 μm, wood chips, plastic waste, and, to a lesser extent municipal waste. For hazardous wastes that are fine powders (50 μm or less in size), suspension firing of the powders is practical. Reduction of solids to the sizes required for suspension firing would not be justified for small-scale units.

B. Stokers

There are several types of stokers designed to feed solids onto a furnace grate. The solids may be fed to the grate through a chute or propelled by an impeller over the grate. The solids burn in the bed with a portion of the air supplied through the grate and the balance through jets above the grate. Motion of solids through the furnace is accomplished by the use of reciprocating mechanical stokers, including vibrating grates, traveling grates, and rocking grates. Stokers have been used to burn coal and a variety of waste products such as wood waste and bagasse; they are used extensively for the disposal of municipal waste.

C. Single- and Multiple-Hearth Incinerators

These incinerators are well suited for the disposal of sludges and solids containing high concentrations of water and/or ash. Within such an incinerator, wastes undergo four successive processes: (1) drying, (2) combustion of the dried waste (a substantial auxiliary fuel requirement is normal), (3) ash cooling, and (4) ash discharge. Although single-hearth incinerators have been used with success, tight control of the drying/combustion/cooling sequence is achieved only by utilizing multiple hearths, typically five to nine hearths. A typical multiple-hearth incinerator system is shown in Figure L.2. The ash residence time in the furnace is quite long, typically 30 minutes or more.

Most modern multiple-hearth incinerators are equipped with a downstream afterburner (typically operating at about 815°C), which destroys hydrocarbon fumes that are only partially oxidized in the upper (waste-drying) hearths. An afterburner is a necessity if the combustion zone of the furnace is operated in the pyrolysis/low-temperature mode to

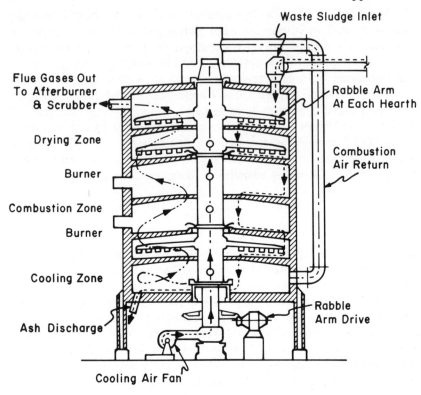

FIGURE L.2   Multiple-hearth incinerator.

prevent ash fusion problems (e.g., many sodium salts melt in the range 590–815°C). Since auxiliary fuel normally is used in both the combustion zone of a multiple-hearth furnace and the afterburner, it is possible to use hydrocarbon wastes to supplement the primary fuels traditionally fired in these burners.

D.   ROTARY KILNS

The rotary kiln has become the preferred type of incinerator for solid wastes because of its ability to handle a wide variety of wastes. Loose waste material is usually fed into the rotary kiln by a hydraulic ram; the waste slowly breaks up and burns or tumbles down the rotating cylindrical chamber (see Figure L.3 for an example). Containerized wastes in cardboard or even steel drums can be charged to the kiln by special ram feeders.

The refractory lining of a rotary kiln can be damaged by mechanical

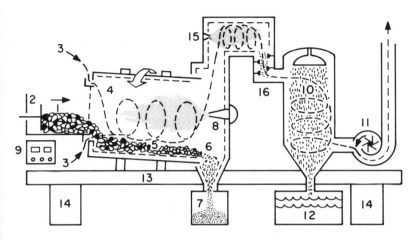

1 Waste to incinerator
2 Auto-cycle feeding system:
   feed hopper, pneumatic feeder, slide gates
3 Combustion air in
4 Refractory-lined, rotating cylinder
5 Tumble-burning action
6 Incombustible ash
7 Ash bin
8 Auto-control Burner Package:
   programmed pilot burner
9 Self-compensating instrumentation-controls
10 Wet Scrubber Package:
   stainless steel, corrosion-free wet
   scrubber; gas quench
11 Exhaust fan and stack
12 Recycle water, fly-ash sludge collector
13 Support frame
14 Support piers
15 Afterburner chamber
16 Precooler

FIGURE L.3 Rotary kiln incinerator.

abuse (steel drums cannot be charged very gently) and by high-temperature corrosion caused by certain ash species. In Europe kiln linings of air-cooled Umco alloy (Co–Cr–Fe) have proven to be superior to refractory linings.

An afterburner is common especially when kiln temperatures are held down (typically 700°C) to minimize corrosion and ash fusion problems in the kiln.

E. FLUIDIZED-BED INCINERATORS

Fluidized-bed incinerators are a logical extension of the fluidized-bed technology that has been developed for roasting, drying, calcining, and catalytic cracking. A solid waste or sludge can be incinerated effectively by injecting it into a fluidized bed of hot, inert particles of sand; the intense mixing of the fluidized bed permits operation of such an incinerator with fairly low bed temperature (typically 700–815°C) and low excess air (3–5% oxygen in the flue gas). A typical unit is illustrated in Figure L.4.

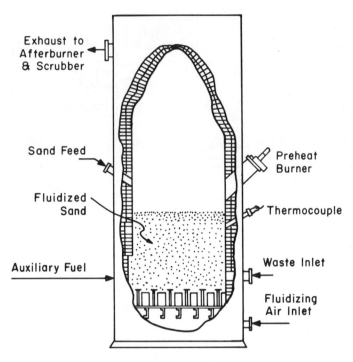

FIGURE L.4   Fluidized-bed incinerator.

If the initial melting temperature of the ash in the waste is less than the fluidized-bed temperature, sticky ash particles may form large ash/sand clinkers in the bed, eventually leading to bed defluidization and a forced shutdown of the incinerator. Addition of kaolin clay and similar high-melting materials to biosludges can increase the initial melting temperature of the ash in the fluidized bed and eliminate this problem.

If the bed temperature is low (below about 650°C), an afterburner must be used to prevent excessive emissions of carbon monoxide and odors.

F.   CYCLONES

Cyclone combustors (vortex burners) operate by tangential injection of combustible and air into a cylindrical chamber. Centrifugal forces drive the solids against the surface of the combustor or keep the solids in circulation until they are small enough to be entrained with the combustion gases. Cyclone burners provide a longer residence time for solids than for gases and can therefore be used to burn solids with particle sizes considerably larger than those suitable for pulverized solid burners. Cyclone burners have been widely used for coal combustion, particularly of low-rank coals, and a variety of waste products.[1]

G. CATALYTIC INCINERATION

The temperatures at which oxidation of organic compounds can be effected may be significantly reduced by passing mixtures of organic materials over a catalyst. Oxidation catalysts consist of noble metals (e.g., platinum, palladium, and rhodium) or oxides of nickel, cobalt, or manganese deposited on a refractory substrate. The catalyst support may consist of particles in a packed or fluidized-bed support, a grid, or a monolithic honeycomb support. The advantages of catalytic incineration are that it can be conducted at relatively low temperatures with low organic concentrations considerably outside of the flammability limits and that high volumetric reaction rates can be achieved. The disadvantages are that the organic compounds must be vaporized and premixed with oxidant, that catalyst cost may be high, and that catalysts may be poisoned by sulfur or other compounds. Catalytic incineration has been used extensively for the destruction of dilute concentrations of hydrocarbons in gas streams. It has also been used for the destruction of chlorinated hydrocarbon waste.[1]

H. MOBILE INCINERATORS

A mobile incinerator mounted on three heavy-duty, air-suspension semi-trailers has been built by the EPA and is undergoing tests.[2] The system is designed to meet the requirements for incineration of polychlorinated biphenyls, with a minimum residence time of 2 seconds at 1200°C, 3% excess oxygen, and a $(CO_2)/(CO_2 + CO)$ ratio of at least 99.9%. It has an estimated capacity of about 4000 kg/h.

Another type of mobile incinerator is one built on an oceangoing vessel. The M/T *Vulcanus* has burned liquid organic wastes containing 64% chlorine at 1200°C with a destruction efficiency greater than 99.9%. This ship's incinerator system consists of two large refractory-lined combustion chambers, fired from the bottom and equipped with a stack on top. It can burn up to 25 metric tons/h. It is not equipped with scrubbers. Pollutants emitted from the stack are either dispersed into the atmosphere (e.g., Cl_2) or absorbed and dumped into seawater (e.g., HCl).

260

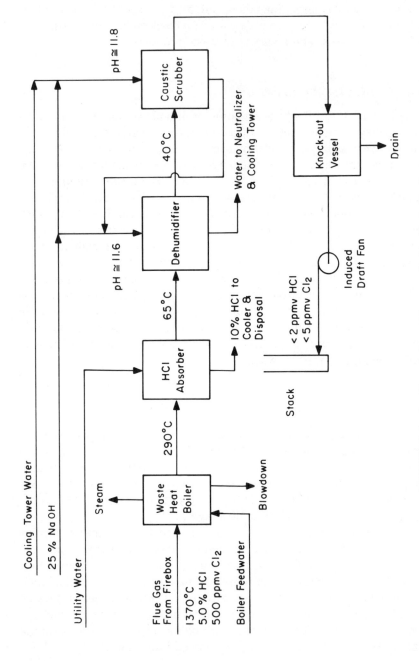

FIGURE L.5   Auxiliary equipment for organic chloride incinerator.

II. AUXILIARY EQUIPMENT

This section describes the principal types of process equipment (except waste-heat recovery devices, which are covered in Section III, below) frequently located downstream from an incinerator firebox. An example of the auxiliary equipment employed in an organic chloride incineration system is shown in Figure L.5.

A. QUENCH

A quench chamber is used to cool the flue gas leaving the firebox (e.g., about 1100°C) to a temperature suitable for the materials used in the downstream flue gas-treatment equipment and stack, typically about 100°C. Cooling is accomplished by direct heat transfer, i.e., injecting cold water or air into the hot flue gas. With water quenches the quench vessel is normally a spray tower; depending on the design, all or part of the spray water is evaporated ("dry bottom" and "wet bottom" designs, respectively). The walls of the water quench tower can be protected by a refractory or by a continuous water curtain.

The option of recovering heat from the hot combustion gases leaving the firebox is lost when a quench is used. Waste-heat recovery should always be considered as an alternative.

B. SCRUBBERS

The principal function of a scrubber is to remove gaseous pollutants (e.g., HCl and SO_2) from cooled combustion gases. A secondary function is to remove particulates (e.g., unburned carbon and flyash). A scrubber is a mass-transfer device in which the flue gas is contacted with a scrubbing liquor (e.g., water), and gaseous pollutant is captured by absorption into and sometimes reaction with the scrubbing liquor.

A wide variety of scrubber designs have been developed, but they can be divided into two basic types—traditional and high energy. Traditional scrubbers, either packed beds or spray towers, are inexpensive and have a low gas pressure drop (13 cm of water column); however, a series of such devices is often required to achieve the desired scrubbing efficiency. High-energy scrubbers, venturi and eductor scrubbers, are compact and can achieve high scrubbing efficiencies at the expense of high gas pressure drop (as high as 150 cm of water column) for a venturi scrubber or high liquid pressure drop (200 or 300 psi) for the sprays in an eductor scrubber.

Scrubbers are used extensively in organic chloride incinerators. HCl is readily absorbed by water in spray towers. On the other hand, the solubility of Cl_2 in water is low, and so caustic scrubbing is used (Cl_2 at the gas/liquid interface reacts with aqueous NaOH to form mainly aqueous NaCl).

A major drawback associated with the use of scrubbers is the need to dispose of spent scrubbing liquor.

C. PARTICULATE COLLECTORS

Particulate collectors remove solid particles and aerosols from cooled flue gas. The three major types are discussed below.

The traditional mechanical devices include the cyclone, baffle, and chevron (zigzag) separators. Gas pressure drop for such collectors is low (13 cm of water column), but the removal efficiency for fine particulates, especially those smaller than ~10 μm, is poor.

High-energy scrubbers are better fine-particle collectors than are mechanical collectors, but even venturi and eductor scrubbers are not very effective in removing particles smaller than 1–2 μm.

The devices capable of good fine-particle removal are the electrostatic precipitator (ESP) and the baghouse. In an ESP a particle is charged in an electric field and then migrates to a collecting surface (e.g., a steel plate) where it loses its charge; collected particles are removed from the collector by rapping or washing. ESP's can operate at temperatures up to about 430°C. A baghouse is an enclosure containing many individual fabric filters shaped like a bag or sock. The dirty side of each bag is cleaned periodically by a shaker or a pulsed jet. Developments in the synthetic fiber industry continue to increase the maximum operating temperature of a baghouse; the current ceiling is about 260°C. High capital and maintenance costs are associated with ESP's and baghouses; they may be too expensive to use in the smaller incineration systems envisaged for individual laboratories.

D. DEHUMIDIFICATION

In some communities the emission of a heavy "steam" plume from an incinerator stack may be undesirable, necessitating the use of a dehumidifier. Dehumidification can be achieved in a spray tower, which typically condenses and removes ~70% of the water vapor from the flue gases.

E. NO_x AND SO_x CONTROL

Emissions of nitrogen and sulfur oxides from incinerators are a source of major concern. A wide assortment of control techniques involving combustion modification and flue-gas treatment has been developed for boilers. Almost all of the NO_x control methods that involve combustion modification (e.g., low excess air) will have an adverse effect on an incinerator's waste-destruction efficiency.

F. DISPOSAL OF ASH AND SCRUBBER LIQUORS

Dry ash from a hazardous-waste incinerator is considered to be a hazardous waste by definition and must be disposed of in a secure landfill. An incinerator operator who can establish that his ash does not meet any of the characteristics of a hazardous waste can apply to the EPA for an exemption that will permit the ash to go to a sanitary landfill if local regulations allow it.

Recycle of scrubber liquors is the best way to reduce the quantities requiring ultimate disposal. These liquors are often disposed of by evaporation in ponds or by injection into deep wells.

III. WASTE-HEAT RECOVERY

There is a strong economic incentive to recover the thermal energy from the flue gas leaving an incinerator firebox at, for example, about 1100°C. This recovery can be accomplished by indirect heat transfer to preheat the combustion air or to generate steam. Typically, cooled flue gas exits a waste-heat exchanger at 260°C to 315°C, and over 70% of the heating value of the wastes and fuels burned in the firebox can be recovered. A heat recovery system is more likely to be worth the investment if it is designed in a new incinerator installation than if it is added to an existing incinerator.

Waste-heat exchangers usually are shell-and-tube units where the combustion gases either pass through the tubes (i.e., fire-tube design) or through the shell side (i.e., water-tube design if steam is being generated). The major problems encountered are fouling of the heat-transfer surfaces and plugging of the gas passages by ash species in the flue gas. The fouling and plugging problems can be managed by limiting the ash content of the wastes or by using devices capable of cleaning the exchanger on line (e.g., soot blowers).

IV. MATERIALS CONSIDERATIONS IN INCINERATION PROCESSES

Reactions of metals and refractory materials with corrosive substances is a complex subject. This summary can only indicate the factors that influence selection of construction materials.

A. INCINERATOR/COMBUSTOR

Gases from waste chemical incinerators contain a greater variety and amount of acidic components than gases from boilers or furnaces. Accordingly, there may be limitations on materials of construction beyond those for the fire side of steam boilers. Components in the combustion gases from waste incinerators may include low concentrations of chlorine, hydrogen chloride, sodium chloride, and oxides of nitrogen, sulfur, phosphorus, sodium, potassium, calcium, silicon, and aluminum.

B. REFRACTORIES

The selection of refractories for incinerators is usually dictated by a number of factors other than resistance to the furnace atmosphere. Primary considerations are likely to include temperature, cost of materials, furnace geometry, thermal conductivity, and resistance to thermal shock and erosion. For the most part refractory brick, or ram plastics with a high alumina content and capable of sustained operation at or near 1540°C, will be selected. For the most severe service, brick is preferred over monolithics, although burner design and geometry may dictate the latter.

C. ALKALIS

If contaminated fuels are used to provide the main source of heat, severe problems may be expected in the burner throat. Bunker C or lower-quality fuels, especially those with high alkali content, burned in high-intensity burners typically limit lifetimes to the order of 3–5 months.

For the main lining of the firebox, brick—most often of superduty grade—is to be preferred, although for many small-vent gas burners high-alumina phosphate-bonded plastics can be used. As with burner throats, attack of the firebox linings by alkali at temperatures greater than about 980°C is a common problem. Damage can often be limited by choosing high-fired 70% alumina materials for the hotter zones.

Other inorganics often found in wastes, such as HCl, HBr, Cl_2, Br_2, and P_2O_5, are not particularly destructive to alumina-silica refractories. The presence of P_2O_5 may cause some problems as a result of condensation of orthophosphoric acid below about 800°C. The presence of SO_2/SO_3 in the vapor has little effect on brick structures, but calcium-aluminate-bonded castables are prone to failure.

Thermal shock damage to refractory linings is often a cause of catastrophic failure. As a general rule, materials of moderate strength and high porosity are relatively resistant to thermal shock.

D. METALLIC MATERIALS OF CONSTRUCTION

Metals and alloys cannot be used as construction materials over such wide temperature ranges as can refractories and, except for large membrane wall-type furnaces used in municipal incinerators, one does not usually find exposed metal in a chemical incinerator firebox. Use of metals is generally limited to temperatures less than about 980°C. Most metals have poor mechanical properties even well below this temperature, and oxidation is significant at this temperature. Reduced tensile properties frequently determine the maximum useful temperature. Uncooled metal parts in incineration systems are atypical except for thermowells and/or boiler tube supports.

E. CORROSION

Compounds of vanadium, sodium, potassium, sulfur, molybdenum, or lead can accelerate corrosion of steels, stainless steels, and other metallic materials of construction at elevated temperature. Some of these constituents are found in fuels and wastes, and while their concentration might be quite low, deposits on metal surfaces in boilers may contain these components in high concentrations. Vanadium and sodium form products melting as low as 500°C.

Carbon steel and low-alloy steels are not suitable for exposure at furnace temperatures in the presence of hydrogen chloride and/or chlorine. Typical maximum temperatures for exposure of carbon steel or low-alloy steels to chlorine is approximately 200°C and for hydrogen chloride, about 260°C. If higher metal temperatures are envisaged, high-nickel alloys are preferred.

If heat recovery is to be incorporated in an incineration system, a large surface area of metal becomes exposed to the flue gas, and considerable thought should be given before such a step is taken. Where halogens or SO_2/SO_3 are present in the gas phase, consideration should be given to incorporating provision for a purge step prior to a shutdown to avoid dewpoint corrosion problems on cool surfaces.

F. QUENCH SECTION

Materials of construction in the quench section may be subject to rapid temperature variations or thermal shock and alternate wetting and drying of exposed surfaces. The quench liquid may be present in some local areas as a highly concentrated aqueous solution, and moist salts may be present on structural surfaces. Acid gases in the flue gas could condense as a liquid phase that is strongly acidic even though the overall quench liquid is only slightly acidic. The quench system can, from a materials standpoint, be divided into two zones—one that is exposed to hot flue gases at 760–1100°C and a transition zone that is exposed to both hot gases and quench liquid.

For the hot zone, brick construction is suitable with standard air- or heat-set mortars. For the transition zone, where hot flue gases and quench liquid are present together, careful materials selection is usually critical to successful operation. While brick linings are typically chosen for this section, an essential requirement is that they be resistant to thermal shock and to the acid or alkaline conditions of the quench liquid.

Metal-lined transition sections are usually not satisfactory unless separately cooled. Horizontal quench systems, where the hot flue gases enter a horizontal quench chamber, are to be avoided because severe problems develop at the entrance zone where alternate wet and dry conditions develop. Impingement of quench sprays on hot refractories is often more the rule than the exception, resulting in severe structural damage to the brick lining. Wherever possible, a vertical quench with a weir or wetted-wall transition section should be used. In the wetted-wall approach, adequate quenching of the hot gases is usually more certain, so that organic materials of construction can be used for the quench vessel.

G. PLASTICS IN SCRUBBERS

Scrubber environments typically contain a variety of chemical species. Aqueous salt solutions and dilute acids do not present a severe chemical environment for many polymers. The components of the scrub liquid detrimental to polymeric materials are ozidizing aqueous solutions of halogens and of their hypoacids. Decreasing the concentration of strongly oxidizing species by chemical means, e.g., addition of sodium sulfite, can extend the useful life of plastic equipment.

If hydrocarbons are present in the scrub liquid, the suitability of each material should be re-evaluated. Some hydrocarbons could result in elimination of many plastic materials from consideration or could require lowering the maximum service temperature.

REFERENCES

1. M. Sittig, *Incineration of Industrial Hazardous Wastes and Sludges,* Pollution Technology Review No. 63, Noyes Data Corporation, Park Ridge, N.J., 1979.
2 J. E. Brugger, J. J. Yezzi, Jr., I. Wilder, F. J. Freestone, R. A. Miller, and C. Pfrommer, Jr., *Proceedings of the 1982 Conference on Hazardous Material Spills,* Milwaukee, Wisc., April 19–22, 1982, p. 116.

M   Sources of Information on Hazardous-Waste Regulations

Information on the regulations that apply to disposal and transport of chemical wastes from laboratories can be obtained from the sources listed below. Complete summaries of hazardous-waste regulations can be obtained from commercial publishers, and information on locating them is available to members of the American Chemical Society (ACS) through the ACS Department of Public Affairs, Office of Federal Regulatory Programs, 1155 16th Street, N.W., Washington, D.C. 20036. Listed below are sources of information on regulations that pertain to waste-disposal and transport and to proposed and actual changes that are continually being made to these regulations.

1. The *Federal Register* (FR) is the official publication of the U.S. government for notifying the public of proposed and final regulations. It provides a uniform system for federal agencies to make regulations and legal notices available to the public. Four basic types of documents can be found in the FR: presidential documents (e.g., proclamations, executive orders, memoranda of understanding), final agency rules, proposed rules, and agency notices. Because proposed regulations are published in the FR for comment before becoming final, the FR is a useful tool for those who would like to participate in the regulatory process through their comments on proposed regulations, which become part of the public record that agencies consider before issuing final rules. The FR is also a valuable tool for keeping up to date on federal regu-

lations in effect. There are no restrictions on reproduction or republication of material in the FR.

The FR is readily available in most large libraries and is published daily, except for Saturdays, Sundays, and official federal holidays, by the U.S. Government Printing Office. An edition of the FR on 24× microfiche is also published daily. Annual editions are available on microfilm. The FR Monthly Index, which is cumulative throughout the year, is part of a FR subscription. It is also available by separate subscription or as single copies. The December issue is the annual index to the FR. Citations to the FR are of the form "39 FR 1248" for an item that begins on page 1248 of the thirty-ninth volume of the *Federal Register*.

2. The *Code of Federal Regulations* (CFR) is the annual codification of all federal regulations. Each volume of the CFR is updated annually, although issue dates for these volumes are staggered throughout the year. Most large libraries own the CFR. Single volumes of the CFR are available for purchase, and information on the CFR printing schedule and prices can be obtained from the CFR Unit of the Office of the Federal Register: (202) 523-3419.

Citations to the CFR are of the form "40 CFR 10.2" for an item in Title 40 of the *Code of Federal Regulations*, part 10, section 2.

The volumes of CFR that are pertinent to this report are:

40 CFR Chapter I, Environmental Protection Agency

49 CFR Chapter I, Transportation

Users of the CFR can remain current with proposed and final amendments published in the FR by consulting the *List of CFR Sections Affected* (LSA), a monthly numerical index prepared by the Office of the Federal Register. The LSA is part of a FR subscription but is also available by separate subscription and as single copies.

A general subject index to the entire CFR appears in the *CFR Index and Finding Aids* volume, which is revised as of 1 January of each year.

The FR, CFR, LSA, Federal Register Index, and CFR Index can be ordered from the Superintendent of Documents, U.S. Government Printing Office, Washington, D.C. 20402. Telephone orders are accepted (payment by MasterCard/Visa), and price information is given by the Government Printing Office Order Desk: (202) 783-3238.

Answers to general information questions and assistance in locating specific items in the FR or CFR can be obtained from the Indexing Unit of the Office of the Federal Register: (202) 523-5227.

3. *Washington Alert* is a bulletin published biweekly by the ACS

Department of Public Affairs, which is available without charge to members of ACS. It includes abstracts of proposed and final federal regulations that may affect chemical laboratories.

4. *State Hazardous-Waste Regulations*. There are a number of commercial publications that summarize or give details of the hazardous-waste regulations of the individual states. A summary of state hazardous-waste laws, *Hazardous Waste Management: A Survey of State Legislation 1982*, can be purchased from the National Conference of State Legislatures, 1125 17th Street, Suite 1500, PR2, Denver, Colo. 80202. Information on the regulations of a specific state can be obtained from the environmental regulatory agency of that state (see Appendix B).

Glossary of Abbreviations
and Selected Terms

ACS	American Chemical Society
CCBW	Chemically contaminated biological waste
CFR	Code of Federal Regulations
DOT	U.S. Department of Transportation
EPA	U.S. Environmental Protection Agency
FR	*Federal Register*
Lab pack	See Chapter 10, Section V.B, for a general description. See Appendix A, 40 CFR 265.316.
LC_{50}	A concentration in air that is lethal to 50% of a group of test animals.
LD_{50}	A dose ingested, injected, or applied to the skin that is lethal to 50% of a group of test animals.
NRC	National Research Council, the operating arm of the National Academy of Sciences and the National Academy of Engineering.

n.o.s.	Not otherwise specified (often used in DOT classifications and regulations).
ORM-A	Other Regulated Material; see Appendix D, 49 CFR 173.500(b)(1).
ORM-B	Other Regulated Material; see Appendix D, 49 CFR 173.500(b)(2).
ORM-E	Other Regulated Material; see Appendix D, 49 CFR 173.500(b)(5).
PCB	Polychlorinated biphenyl; EPA interprets this term to include the monochlorobiphenyls.
POHC	Principal Organic Hazardous Constituent: an organic chemical that is a constituent that is to be burned in an incinerator and that has been identified by EPA in Appendix VIII of 40 CFR Part 261.
Poison A	See Appendix D, 49 CFR 173.326.
Poison B	See Appendix D, 49 CFR 173.343.
POTW	Publicly Owned Treatment Works; designates a municipal wastewater treatment plant.
RCRA	Resource Conservation and Recovery Act
Secure landfill	A landfill that is authorized by EPA or a state to receive hazardous waste.
WWTP	Wastewater treatment plant.

Index